INTERNATIONAL CENTRE FOR MECHANICAL SCIENCES

COURSES AND LECTURES - No. 74

PETER GLANSDORFF
UNIVERSITÉ LIBRE, BRUXELLES

THERMODYNAMICS IN CONTEMPORARY DYNAMICS

LECTURES HELD AT THE DEPARTMENT
OF MECHANICS OF SOLIDS
JULY 1971

UDINE 1971

SPRINGER-VERLAG WIEN GMBH

Originally published by Springer-Verlag Wien-New York in 1972

ISBN 978-3-211-81177-1 ISBN 978-3-7091-2844-2 (eBook)

DOI 10.1007/978-3-7091-2844-2

P R E F A C E

The notes which follow correspond to the 12 lectures given at CISM in Udine, during the summer course of 1971. The first part represents a summary of the monograph published with my colleague PRIGOGINE under the title: "Thermodynamics theory of Structure, Stability and Fluctuations" (G. Wiley Interscience, London, N.Y., 1971). A french edition of the same monograph has been published about the same time in Paris, by MASSON ET CIE, under the title of "Structure, Stabilité et Fluctuations". This summary should enable the reader to grasp the interest for a non linear thermodynamics of irreversible phenomena, i.e. those who occur far from equilibrium. Indeed, it is within this range that instabilities can be expected, which in turn are susceptible to give birth to some space or time order in the system (limit cycle, etc...). Several examples illustrate this important generalization of classical thermodynamics, as well as its extension to hydrodynamics, by taking into account the convective effects along with dissipative phenomena, usually considered alone.

The second part is devoted to original new contributions which are not included in the above monograph. In this respect, let us mention on the theoretical side, the demonstration of the convergence of the iteration method used together with the local potential, in order to determine approximate solutions when the exact solution cannot be expected. On the other hand, from the standpoint of practical examples, let us underline the next application treated in Chap. XI, and devoted to an extension of the Bénard problem, to a two component system. Numerous graphs and numerical data due to Dr. J.C. LEGROS are therein established.

Udine, July 1971

1. Introduction.

In the notes which follow, the reader will distinguish two parts:

1°) Sections which constitute a short summary of the laws of non linear thermodynamics, extensively developed in a recent monograph (P. Glansdorff and I. Prigogine, "Thermodynamic Theory of Structure Stability and Fluctuations", J. Wiley, Interscience, London, N.Y. 1971, or the French edition: "Structure, Stabilité et Fluctuations", Masson, Paris, 1971).

2°) Sections which present a detailed account of new applications. The most important is related to the stability of a two component fluid layer, heated from below. Numerous numerical results obtained by J.C. Legras are given in Section 11.

Using the "local equilibrium" assumption valid for a great variety of problems in macroscopic physics, one can develop a more general thermodynamics of irreversible processes including the nonlinear domain as well. An important consequence of this extension is to bring into light the instabilities and the associated critical states at which the system passes from one "branch" to another.

This new thermodynamics has to comprise not only the statement of the system's behavior, but also that of its fluctuations. The neighbourhood of equilibrium appears then as a region of regressing fluctuations which means that the stability conditions are identically satisfied therein. On the contrary the far from equilibrium region may admit the growth of the latter, generating instabilities and marginal states, thus bringing about new solutions situated on other branches. This new macroscopic presentation of irreversible phenomena makes conspicuous the following duality in thermodynamics:

- In the region of linear laws, i.e. for weak constraints, thermodynamics preserves its well known characteristic of a science related to destructions of structure.
- In the far from equilibrium region, i.e. for strong constraints, instabilities can be generated, enabling us to interpret thermodynamics as a science including the elaboration of new structures.

These are the so-called dissipative structures.

Several examples are then treated to analyse the fundamental mechanism, giving rise to instabilities which lead in turn to such structure. So far auto-catalysis appears amongst the most important mechanism generating chemical as well as biological instabilities.

Other examples illustrate a second extension, to take the convective effects into account along with dissipative ones, since both influence directly the stability. The following account presents therefore a macroscopic physics providing us with the same general method to approach the hydrodynamic, thermal, chemical and biological stability problems.

2. Balance Equations – The Conservation Laws.

Let us consider an extensive variable defined by the volume integral (*)

$$I(t) = \int f(x_i, t)\, dV \tag{2.1}$$

Assuming the boundary surface Ω at rest, we follow the time change of (2.1). This may be split into two parts:

$$dI = d_e I + d_i I \tag{2.2}$$

where $d_e I$ denotes the variation due to the exchange with the external world, and $d_i I$ the source of the quantity I inside the system. A conservation law for the variable I implies:

$$d_i I = 0 \quad \text{i.e.} \quad dI = d_e I \tag{2.3}$$

As an example, let us consider the first law of thermodynamics, which expresses the conservation of energy $(I = E)$.
In a closed system and in the absence of an external field of volumic forces, the exchange $d_e E$ corresponds to the heat flux dQ and to the mechanical work dW performed by the stresses on the boundary of the system. If the stresses reduce to an uniform pressure, normal to the surface, the work is simply $-p\,dV$ and the conserva-

(*) The limit V is not indicated explicitly.

tion law becomes:

$$d_i E = 0 \quad \text{or} \quad dE = dQ - p dV \tag{2.4}$$

Likewise, the change of mass of some component γ of the system involves an external exchange $d_e m_\gamma$ with the outside world as well as a source $d_i m_\gamma$ due to the chemical reactions. One has:

$$dm_\gamma = d_e m_\gamma + d_i m_\gamma \tag{2.5}$$

For instance, for a single chemical reaction we may write

$$d_i m_\gamma = \nu_\gamma M_\gamma \, d\xi \tag{2.6}$$

where M_γ denotes the molar mass of γ , and ν_γ its stoechiometric coefficient in the chemical reaction. The quantity ξ is the chemical variable, i.e. the degree of advancement of the chemical reaction considered, and

$$w = d\xi / dt \tag{2.7}$$

is the chemical reaction rate.

Going back to the general relation (2.1), (2.2) we may also derive the corresponding balance equations as

$$\frac{\partial f}{\partial t} = - \nabla \mathbf{j}[I] + \sigma[I] \tag{2.8}$$

As a rule, the flow $\mathbf{j}$ associated to I , contains a conduction flow and a convection flow. On the other hand $\sigma[I]$ is the local source defined as:

$$P[I] = \frac{d_i I}{dt} = \int \sigma [I] \, dV \tag{2.9}$$

The general expressions of the local balance equation for mass, momentum and energy are to be found in the monograph [1].

3. The Second Law of Thermodynamics and the Local Equilibrium Assumption.

As well known, the entropy is the suitable extensive variable to express the second law of thermodynamics. Writing the entropy balance equation as:

$$dS = d_e S + d_i S \tag{3.1}$$

the second law prescribes a positive value to the entropy production, whatever the irreversible process inside the system may be. In other words, the inequality:

$$d_i S \geqslant 0 \tag{3.2}$$

vanishing for a reversible change (equilibrium process), is always valid.

The explicit expression of the entropy production in terms of the variables generally used to describe the state of a macroscopic system, as e.g. temperature, pressure, mass fraction, ..., is derived in what follows using the so-called local equilibrium assumption.

This means that the local specific entropy s is related to the above variables by the same expression as in equilibrium. This is the well known Gibbs relation:

$$T\delta s = \delta e + p\,\delta v - \sum_{\gamma} \mu_{\gamma}\,\delta N_{\gamma}\,; \qquad (v = \rho^{-1};\ \Sigma N_{\gamma} = 1) \tag{3.3}$$

The variables T, p and the chemical potentials μ_{γ} (per unit mass) have the same definition here, as in equilibrium thermodynamics. The δ denotes an arbitrary small increment. Then, the mass and the energy balance equations, used together with (3.3), give rise directly to the explicit form sought of the local entropy balance equation. The general formalism (2.8) yields here:

$$\sigma[s] = \partial_t(\rho s) + \nabla \mathscr{S} \tag{3.4}$$

where the entropy production per unit mass is explicitly (see the Glossary for the non selfexplanatory notations) [1]:

$$\begin{aligned} \sigma[s] = {} & W_j T^{-1}_{,j} - \sum_{\gamma} (\rho\,\Delta_{\gamma j})\left[(\mu_{\gamma} T^{-1})_{,j} - T^{-1} F_{\gamma j}\right] \\ & - p_{ij} T^{-1} v_{i,j} + \sum_{\rho} w_{\rho} \cdot A_{\rho} T^{-1} \geqslant 0 \end{aligned} \tag{3.5}$$

$$(\rho = 1,\dots, r;\ i, j = x, y, z;\ \partial/\partial x_j \equiv {}_{,j}).$$

or in the more compact bilinear form:

$$\sigma\left[S\right] = \sum_{\alpha} J_{\alpha} X_{\alpha} \tag{3.6}$$

using generalized flows J_{α} and generalized forces X_{α}. Then, according to (2.9), we get for the total entropy production per unit time

$$\frac{d_i S}{dt} = P\left[S\right] = \int \sigma\left[S\right] dV \tag{3.7}$$

On the other hand, the entropy flow is given by:

$$\boldsymbol{\mathcal{S}} = \mathbf{W} T^{-1} - \sum_{\gamma} (\rho_{\gamma} \boldsymbol{\Delta}_{\gamma})(\mu_{\gamma} T^{-1}) + \rho s \mathbf{v} \tag{3.8}$$

The splitting into a source term and a flow term in (3.4) has been performed in such a way that $\sigma\left[S\right]$ vanishes at equilibrium. Moreover one observes that the two first terms in the r.h.s. of (3.8) are conduction flows whereas the last term represents a convection flow.

In particular for the simple heat conduction problem, one has:

$$\mathbf{J} = \mathbf{W} \quad \text{(heat flow)} \tag{3.9}$$

$$\mathbf{X} = \nabla T^{-1} \quad (T = \text{absolute temperature}) \tag{3.10}$$

while for a chemical reaction we have to write [cf (2.6),

(2.7)] :

$$J = w \tag{3.11}$$

$$X = -\frac{1}{T}\sum_{\gamma}\nu_{\gamma}\mu_{\gamma} = \frac{A}{T} \tag{3.12}$$

Here μ_γ denotes the chemical potential and

$$A = -\sum_{\gamma}\nu_{\gamma}\mu_{\gamma} \tag{3.13}$$

the chemical affinity.

At equilibrium, both the flows and the forces vanish:

$$J_\alpha = 0 \quad ; \quad X_\alpha = 0 \tag{3.14}$$

This implies that, according to (3.6), the entropy production occurs as a second order quantity in the neighborhood of an equilibrium state.

Non-equilibrium thermodynamics concerns situations in which the equalities (3.14) are no longer valid, but they may still be described macroscopically. Moreover, all the classical thermodynamic relations deduced from the Gibbs formula remain valid and will be sometimes useful in what follows. Most of them are well known and can be found in various textbooks on thermodynamics. Therefore these will be no longer reproduced hereafter. However, we shall establish separately the non usual explicit expression of the second differential of entropy, as this quantity plays a basic role in the subsequent approach to the stability problem of an equilibrium or non-equilibrium state.

Taking e, v, N_γ, as independent variables, we derive directly from the Gibbs-relation (3.3), the second-order equality:

$$\delta^2 s = \delta T^{-1} \delta e + \delta(pT^{-1}) \delta v - \sum_\gamma \delta(\mu_\gamma T^{-1}) \delta N_\gamma \qquad (3.15)$$

or, using once more (3.3):

$$T\delta^2 s = - \delta T \delta s + \delta p \delta v - \sum_\gamma \delta \mu_\gamma \delta N_\gamma \qquad (3.16)$$

Now expanding $\delta \mu_\gamma$ in terms of T, p, N_γ, we get after simple manipulations:

$$\delta^2 s = - \frac{1}{T} \left[\frac{c_v}{T}(\delta T)^2 + \frac{\rho}{\chi}(\delta v)^2_{N_\gamma} + \sum_{\gamma\gamma'} \mu_{\gamma\gamma'} \delta N_\gamma \delta N_{\gamma'} \right] \qquad (3.17)$$

One sees here, that the second differential of the specific entropy occurs as a definite quadratic form, in terms of the quantities $\delta T, (\delta v)_{N_\gamma}$, and δN_γ, in the space of the independent variables: e, v, N_γ, as assumed above. This last point has to be kept in mind each time one deals with a second differential.

Let us precise that $(\delta v)_{N_\gamma}$ stands for the differential expression:

$$(\delta v/\delta T)_{p,N_\gamma} \delta T + (\delta v/\delta p)_{T,N_\gamma} \delta p .$$

The main interest of the explicit form (3.17), is to relate the sign of $\delta^2 s$ to those of the coefficients.

Besides, the equality (cf [1]):

$$v\,\delta^2(\rho\, s) = \delta^2 s \tag{3.18}$$

binding $\delta^2(\rho s)$ taken in the ρe, ρ_γ space to $\delta^2 s$ taken in the e, v, N_γ space, is also often useful as it permits the introduction of volumic variables ρe, $\rho_\gamma = \rho N_\gamma$, instead of the corresponding mass variables e, v, N_γ and vice-versa.

The generalization of eqs. (3.15) to (3.18), for complex values of the increments (as it happens in eigenvalue problems), is easily derived by using the substitution:

$$\delta A\,\delta B \rightarrow \frac{1}{2}\left[\delta A\,(\delta B)^* + \delta B\,(\delta A)^*\right] \tag{3.19}$$

where the asterisk denotes the complex conjugate.

The quantity $\delta^2_m s$ such obtained appears as a real definite quadratic expression which reduced to $\delta^2 s$, in case of real increments (for details, see [1]).

4. Linear Thermodynamics and Phenomenological Coefficients.

It results from the equilibrium equalities (3.14), that in its neighbourhood, the link between the flows J_α and the forces X_α may be assumed to be linear:

$$J_\alpha = \sum_\beta L_{\alpha\beta} X_\beta \qquad \alpha,\beta = 1,..,n \tag{4.1}$$

Such relations are called phenomenological laws, where the $L_{\alpha\beta}$ denote the phenomenological coefficients. They include many empirical laws such as the Fourier's law for the heat conduction, the Fick's law for the diffusion, the Newton law for viscous fluids, and so on. The entropy production (3.6) takes here the quadratic form:

$$\sigma[S] = \sum_{\alpha} \sum_{\beta} L_{\alpha\beta} X_{\alpha} X_{\beta} \geq 0 \qquad (4.2)$$

implying the inequalities:

$$L_{\alpha\alpha} > 0 \qquad (L_{\alpha\beta} + L_{\beta\alpha})^2 < 4 L_{\alpha\alpha} L_{\beta\beta} \qquad (4.3)$$

Hence, the proper coefficients $L_{\alpha\alpha}$, $L_{\beta\beta}$, have to be positive quantities. On the contrary the relations (4.1) are rarely useful in representing chemical kinetics, which belongs generally to the non-linear thermodynamics.

Let us observe that in the range of linear problems the phenomenological coefficients have to be considered as pure constants. Otherwise (4.1) is no longer linear. Besides, it must be kept in mind even then, the problem is not always located in the region of linear thermodynamics. Indeed, non-linearity may also arise from the balance equations themselves.

An important property of the coefficients $L_{\alpha\beta}$ is given by the so-called Onsager reciprocity relations, that is

$$L_{\alpha\beta} = L_{\beta\alpha} \qquad (4.4)$$

This means that J_{α} depends on the force X_{β} of the irreversible process β, as J_{β} depends on X_{α}. The demonstration is based on the principle of microscopic reversibility, but is out of the scope of the present statement. In some cases (presence of magnetic field), this principle is no longer valid.

In the linear range, some irreversible processes may be coupled through a mixed coefficient $L_{\alpha\beta}$ ($\alpha \neq \beta$), but others cannot. The vector character of the process plays here an important role.

Coupling is only possible between quantities having the same tensor character. This property has been connected by Prigogine (1947) to the Curie theorems on symmetry in crystal physics. For instance chemical reactions are described by scalars, heat flow and diffusion flow by vectors, momentum flow by a second order tensor.

However for anisotropic medium, this character may be modified. For instance, chemical reactions inside an anisotropic membrane have to be described sometimes by vector quantities, and this allows couplings with diffusion processes [active transport, cf. Katchalsky (1968)] .

Let us finally consider (with Prigogine [1965]) the Electrokinetic effect as an application using the Onsager's relations.

Two vessels are separated by a porous wall. The two phases differ only by the pressure and the electrical potential. The entropy production (3.6) may be written here as:

$$P[S] = \frac{J\Delta p}{T} + \frac{I\Delta\varphi}{T} \qquad (4.5)$$

where J and I denote respectively the flow of matter and electrical current; Δp is the pressure difference and $\Delta\varphi$ the electrical potential difference. Equations (4.1) now become

$$I = L_{11}\frac{\Delta\varphi}{T} + L_{12}\frac{\Delta p}{T}$$
$$J = L_{21}\frac{\Delta\varphi}{T} + L_{22}\frac{\Delta p}{T} \qquad (4.6)$$

The following effects may be investigated experimentally:

$(\Delta\varphi/\Delta p)_{I=0} = -L_{12}/L_{11}$ (Streaming potential)

$(J/I)_{\Delta p=0} = L_{21}/L_{11}$ (Electro-osmosis)

$(\Delta p/\Delta\varphi)_{J=0} = -L_{21}/L_{22}$ (Electro-osmotic Pressure)

$(I/J)_{\Delta\varphi=0} = L_{12}/L_{22}$ (Streaming Current)

These four effects may be studied separately and the Onsager formula $L_{12} = L_{21}$, tested experimentally (Mazur and Overbeek, 1951). Many other examples have also confirmed the validity of the reciprocity relations.

5. Stability of Equilibrium States.

Let us expand the entropy S up to the second order as:

(5.1) $$S = S_e + (\delta S)_e + \frac{1}{2}(\delta^2 S)_e$$

This implies around an equilibrium state:

(5.2) $$d_t S = d_t(\delta S)_e + \frac{1}{2} d_t(\delta^2 S)_e$$

since S_e is a time independent quantity.

Accordingly, the global entropy balance equation derived from the local form (3.4) may be split into two relations between first order terms and second order terms. The latter involves the entropy production as observed after eq. (3.14), and yields:

(5.3) $$\frac{1}{2} d_t(\delta^2 S)_e = P[S] - \Delta\Phi[S]$$

The first expresses simply an equilibrium condition.

We assume that well defined boundary conditions are prescribed on the surface of the system, that is:

(5.4) $$\Delta\Phi[S] = \int \left[W . \Delta T^{-1} - \sum_\gamma \rho_\gamma \Delta_\gamma(\mu_\gamma T^{-1}) \right] n \, d\Omega = 0$$

either because the increments ΔT^{-1}, $\Delta_\gamma(\mu_\gamma T^{-1})$, vanish on the boundary, or because the heat flow and diffusion flow vanish through the boundary surface. Then the entropy balance equation reduces to:

$$\frac{1}{2}\, d_t(\delta^2 S)_e = P[S] \geqslant 0 \qquad (5.5)$$

Using now the basic principle introduced by Gibbs and Duhem to express the stability of an equilibrium state, let us prescribe that all arbitrary deviation, starting from the state of interest, cannot be a natural process, as it would contradict the second law of thermodynamics. Therefore the system remains in its equilibrium state and is stable. This stability condition may be written as:

$$\int_e^f P[S]\, dt < 0 \qquad (5.6)$$

or after integration of (5.5):

$$\frac{1}{2}(\delta^2 S)_e = \int_0^t P[S]\, dt = \int_0^t d_i S = \Delta_i S < 0 \qquad (5.7)$$

Going back to the explicit expression of $\delta^2 S$ arising from (3.17), we may derive directly the stability conditions of an equilibrium state in the well known classical form (cf. [1]):

$$c_v > 0 \qquad \text{(thermal stability)} \qquad (5.8)$$

$$\chi = -(\partial v/\partial p)_T / v > 0 \qquad \text{(mechanical stability)} \qquad (5.9)$$

$$(5.10)\qquad \Sigma\ \mu_{\gamma\gamma'}\,\delta N_{\gamma}\,\delta N_{\gamma'} > 0 \qquad (\text{stability with respect to diffusion})$$

The latter includes the chemical stability according to a Jouguet theorem (cf. [1]).

Let us emphasize, that the foregoing statement is by no means that of the pionners Gibbs and Duhem. It is rather, a slight improvement of a procedure initially introduced by Prigo gine on the basis of the properties of the entropy production.

The Gibbs-Duhem line of reasoning is based on a formulation of the second law, valid for closed systems at uniform pressure and temperature, namely:

$$(5.11)\qquad TdS - dE - pdV \geqslant 0$$

Starting from this inequality, the stability condition for an equilibrium state may be written as:

$$(5.12)\qquad \delta S < 0 \qquad (E, V, \text{constant})$$

or alternatively, according to (5.1):

$$(5.13)\qquad (\delta^2 S)_e < 0 \qquad (E, V, \text{constant})$$

By this way, similar formulations are readily derived by simple change of variables. Instead of (5.12), one such obtain

$$\delta H \geqslant 0 \quad \text{(enthalpy)} \qquad (S, p \text{ constant})$$

$$\delta F \geqslant 0 \quad \begin{pmatrix}\text{Helmholtz} \\ \text{Free-Energy}\end{pmatrix} \qquad (T, V \text{ constant}) \tag{5.14}$$

$$\delta G \geqslant 0 \quad \begin{pmatrix}\text{Gibbs} \\ \text{Free-Energy}\end{pmatrix} \qquad (T, p \text{ constant})$$

which also lead to the classical criteria (5.8), (5.9) and (5.10).

The comparison between the two foregoing statements, based both on the definition of the stability, adopted by Gibbs and Duhem, gives rise to the conclusion:

1- The Gibbs-Duhem method provides us with stability criterions valid for small as well as large deviations. Unfortunately the restriction that E and V are to remain constant for the validity of (5.13)-or S, p; T, V; T, p; for conditions (5.14)- does not permit to extend the procedure to the study of stability conditions around a non-equilibrium state, as e.g. a steady state.

2- The preceding method based on the properties of the entropy production provides us with stability conditions valid only for small perturbations (linear stability).
As shown by inequality (5.7), the restriction E, V constant included in (5.13) is ruled out and replaced by the usual type of constraints given in the form of boundary conditions (5.4).

This provides us with a procedure more appropriated to approach the stability problem for non-equilibrium states, at least in the case of small increments (linear stability). This problem is considered in the next section. However, some slight modifications of the statement are yet necessary, due to the generality which imply to derive the excess entropy balance directly form the perturbation equations rather than from the entropy balance equation as above. Finally, let us also stress that hypothesis such as those included in the criteria (5.14) are often inconsistent with the very existence of fluctuations.

6. Stability of Non Equilibrium States.

Let us now investigate the stability of a non-equilibrium state, either steady or unsteady. We shall limit ourselves to small perturbations (linear stability theory). In the present case, we admit from outset, the inequalities (5.8) - (5.10). Therefore, according to (3.17), $\delta^2 s$ occurs either as a negative quadratic form, or as a definite negative quadratic form in the case of purely dissipative systems.

On the other hand, the relation (5.5) is no longer identically satisfied around a non-equilibrium state, and we have now to study the time evolution of $\delta^2 S$, under the constraints prescribed to the system. This suggests to approach the stability problem of the reference state, by the so-called

Liapounoff method, and to adopt $\delta^2 s$ as the suitable Liapounoff function for each point of a dissipative system. This gives rise to the global stability criterion (for details see [1]):

$$\delta^2 S < 0 \tag{6.1}$$

together with

$$\partial_t \delta^2 S \geqslant 0 \quad (t > t_0) \tag{6.2}$$

where the equality sign has to be ruled out for the asymptotic stability condition. Besides the asymptotic condition implies that the l.h.s. of (6.2) occurs as a definite form. The time derivative is taken for constant values of the coefficients.

For an evolution involving convective effects in addition to the dissipative effects above considered (as e.g. hydrodynamic processes), the independent variables e, v, N_γ, used till now to characterize at each point the state of the system are to be replaced by the enlarged set e, v, N_γ, v_i. Then $\delta^2 S$ becomes semi-definite and the definite Liapounoff function to consider is now the negative definite quadratic expression:

$$\delta^2 z = \delta^2 s - T_0^{-1} (\delta \mathbf{v})^2 \quad (z = s - T_0^{-1} \mathbf{v}^2) \tag{6.3}$$

where $\mathbf{v}$ denotes the barycentric velocity, and T_0, the non-varied temperature of the reference state. Putting:

$$Z = \int \rho z \, dV \tag{6.4}$$

the global stability condition (6.1), (6.2) takes here the more general form (see [1]):

(6.5) $$\delta^2 Z < 0$$

together with:

(6.6) $$\partial_t \delta^2 Z \geqslant 0 \qquad (t > t_0)$$

Of course, other Liapounoff functions than $\delta^2 z$ may be used to express the stability condition. However, with $\delta^2 z$, the explicit form of the stability condition is readily obtained. Indeed, taking the time derivative of eq. (3.15.) and using (3.18), (6.3) as well as Euler's theorem on homogeneous functions of the second degree we get (*):

(6.7) $$\begin{aligned} \frac{1}{2}\partial_t \delta^2(\rho z) &= \delta T^{-1} \partial_t \delta(\rho e) \\ &- \sum_\gamma \delta(\mu_\gamma T^{-1} - \frac{1}{2} T_0^{-1} v^2)\, \partial_t \delta \rho_\gamma \\ &- T^{-1} \delta v_i\, \partial_t \delta(\rho v_i) \end{aligned}$$

One sees that the time derivatives included in the r.h.s. are given directly and explicitly by the excess balance equations for mass, momentum and energy, related to the perturbed

(*) As prescribed above, the time derivative is taken for constant values of the coefficients.

motion, that is:

$$\partial_t \delta \rho_\gamma = \sum_\rho \nu_{\gamma\rho} M_\gamma \delta w_\rho - \left[\delta(\rho_\gamma \Delta_{\gamma j} + \rho_\gamma v_j)\right]_{,j} \qquad (6.8)$$

$$\partial_t \delta \rho = -\left[\delta(\rho v_j)\right]_{,j} \qquad (6.9)$$

$$\partial_t \delta(\rho v_i) = F_i \delta\rho - \left[\delta P_{ij} + \delta(\rho v_i v_j)\right]_{,j} \qquad (\delta F_i = 0) \qquad (6.10)$$

$$\begin{aligned} \partial_t \delta(\rho e) &= \sum_\gamma F_{\gamma j} \delta(\rho_\gamma \Delta_{\gamma j}) - \delta(P_{ij} v_{i,j}) \\ &- \left[\delta W_j + \delta(\rho e v_j)\right]_{,j} \qquad (\delta F_{\gamma i} = 0) \end{aligned} \qquad (6.11)$$

Replacing (6.8) - (6.11) into (6.7), we obtain after some manipulations the <u>excess entropy balance equation</u> for small perturbations in the form:

$$\begin{aligned} \frac{1}{2}\partial_t \delta^2(\rho s) &= \sigma[\delta Z] - \Big\{\delta W_j \delta T^{-1} - \sum_\gamma \delta(\rho_\gamma \Delta_{\gamma j})\delta(\mu_\gamma T^{-1}) \\ &- T^{-1}\left[\delta P_{ij}\delta v_i + \frac{1}{2}\rho v_j (\delta \mathbf{v})^2\right] + v_j \delta^2(\rho s)\Big\}_{,j} \end{aligned} \qquad (6.12)$$

where the r.h.s. contains the <u>excess source term</u>:

$$\sigma[\delta Z] = \sum_\alpha \delta J_\alpha \delta X_\alpha$$

$$-\left[\delta(\rho e)\delta v_j + \delta P_{ij}\delta v_i + \frac{1}{2}\rho v_j (\delta \mathbf{v})^2\right] T^{-1}_{,j}$$

$$- \sum_\gamma \delta \rho_\gamma \, \delta v_j \left[F_{\gamma j} T^{-1} - (\mu_\gamma T^{-1})_{,j} \right]$$

$$- \left[\delta P_{ij} \, \delta T^{-1} - T^{-1} \delta(\rho v_j) \, \delta v_i \right] v_{i,j}$$

(6.13)

$$+ \frac{1}{2} T^{-1} (\delta v)^2 (\rho v_j)_{,j}$$

$$+ \left[\delta(\rho e) \, \delta T^{-1}_{,j} - \sum_\gamma \delta \rho_\gamma \, \delta(\mu_\gamma T^{-1})_{,j} \right] v_j^0$$

and the excess flow term between braces.

For instance, in case of unperturbed boundary conditions, the explicit form of the stability condition for a purely dissipative system, reduces to the inequality:

(6.14)
$$P[\delta S] = \int \sum_\alpha \delta J_\alpha \, \delta X_\alpha \, dV \geqslant 0 \qquad (t > t_0)$$

Around an equilibrium state one has $P[\delta S] = P[S]$ and therefore (6.14) is identically satisfied as it was to be expected [cf. (3.6), (3.7)].

Likewise for a non equilibrium state in the range of linear phenomenological laws ($L_{\alpha\beta}$ = constant, see section 4), one has:

(6.15)
$$\sum_\alpha \delta J_\alpha \, \delta X_\alpha = \sum_{\alpha\beta} L_{\alpha\beta} \, \delta X_\alpha \, \delta X_\beta > 0$$

since the r.h.s. is a quadratic form of the same type as the entropy production (4.2) itself.

On the contrary, in the range of non-linear laws, the criterion (6.14) is no longer identically satisfied, and instabilities may be generated in the system for sufficiently large values of the constraints (critical or marginal state), giving rise to the appearance of new branches far from equilibrium (dissipative structures, multiple steady states, ... [1]). As a result the appearance of an instability in a purely dissipative system $(v_i = \delta v_i = 0)$, corresponds to a vanishing value of the excess entropy production $P[\delta S]$.

On the other hand, the relation (6.13) shows also that in the presence of convective effects $(v_i \neq 0)$, the stability condition involves many new contributions, leading generally to great difficulties for the determination of a marginal state. Already for a simple laminar flow, this marginal state which corresponds to the onset of the turbulent motion (critical Reynolds number; see below section 13), can be determined only by approximate methods of numerical analysis.

Another reason to adopt $\delta^2 s$ as the suitable Liapounoff function characterising the stability of a dissipative system, is the possibility to relate the stability theory for small perturbations, to the probability of a fluctuation, by an extension to non-equilibrium states of the Einstein's fluctuation theory. This question is briefly discussed in section 8, with regard to the heat conduction problem.

For complex values of $\delta\rho_\gamma$ $\delta(\rho v_i)$, $\delta(\rho e)$, as it occurs in the stability problem of a single normal mode, solution of the perturbation equations derived from (6.8) - (6.11), the stability criterion (6.5), (6.6) has to be replaced by the inequalities:

(6.16) $$\delta_m^2 Z < 0$$

together with

(6.17) $$\partial_t \delta_m^2 Z \geqslant 0 \qquad (t > t_0)$$

according to the definition of $\delta_m^2 Z$ linked to those of $\delta_m^2 S$ introduced at the end of section 3.

The time dependence of a normal mode is of the form $\exp \omega t$ where ω is in general a complex value $\omega_r + i\,\omega_i$. Therefore one obtains directly after elementary manipulations the equality (see [1]):

(6.18) $$\partial_t \delta_m^2 (\rho\, z) = 2\,\omega_r\, \delta_m^2 (\rho\, z)$$

As the kinetic condition of stability may be written as $\omega_r < 0$, one sees that the relation (6.18) establishes a complete consistency between this kinetic condition and the thermodynamic condition (6.16), (6.17), for each separate normal mode. Therefore these two relations, which usually express a <u>sufficient condition</u> of stability, according to the Liapounoff theory, become a necessary and suffcient condition when applied to each normal mode taken separately.

Let us also emphasize that, contrarily to the kinetic condition, the explicit form of the thermodynamic condition - as e.g. (6.14) for dissipative systems - provides us with the basic mechanism giving rise to an instability.

7. The General Evolution Criterion.

The foregoing stability criteria based on the behaviour of the system close to a given reference state, either steady or unsteady, are only particular cases of the more general stability conditions involving finite deviations as well. Likewise, the latter may be interpreted as arising from some evolution criterion still more general.

For instance, the increase of the entropy with time, i.e. $dS > 0$, corresponds to such a criterion, but its validity is strictly limited to isolated systems, as prescribed by the second law of thermodynamics.

Lately, an appreciable extension, valid for closed and open systems alike, has been derived, under the unique restriction that the prescribed constraints be time independent boundary conditions [1].

Starting from the bilinear expression of the entropy production:

(7.1)
$$P = \int \sum_{\alpha} J_{\alpha} X_{\alpha} \, dV \geqslant 0$$

in terms of the generalized fluxes J_{α} and forces X_{α}, its time change may be written as

(7.2)
$$dP = d_X P + d_J P$$

where

(7.3)
$$d_X P = \int \sum_{\alpha} J_{\alpha} \, d_t X_{\alpha} \, dV$$

(7.4)
$$d_J P = \int \sum_{\alpha} X_{\alpha} \, d_t J_{\alpha} \, dV$$

Then, for a macroscopic system, submitted to time independent constraints, and involving purely dissipative effects, i.e. without mechanical convection, all the natural processes satisfy the general evolution criterion:

(7.5)
$$d_X P \leqslant 0$$

Here the equality sign corresponds not only to equilibrium, but also to steady states. This criterion expresses that the changes of the forces is always such as to <u>decrease</u> the entropy production. It occurs as a differential expression which is not generally an exact differential.

However in the strictly linear range of phenomenological laws, that is when:

$$J_\alpha = \sum_\beta L_{\alpha\beta} X_\beta \qquad (7.6)$$

where the phenomenological coefficients are considered as pure constants satisfying the Onsager reciprocity relations ($L_{\alpha\beta} = L_{\beta\alpha}$), $d_X P$ reduces to an exact differential, according to the equalities:

$$d_X P = d_J P = \frac{1}{2} d P \leqslant 0 \qquad (7.7)$$

easily derived from (7.3) and (7.6). One thus recovers the theorem of the <u>minimum entropy production</u> at steady-state, together with its region of validity, i.e. the linear thermodynamics.

As an example, let us prove the evolution criterion (7.5) for the heat conduction problem in a rigid medium, either isotropic or non isotropic ($\mathbf{J} = \mathbf{W}$; $\mathbf{X} = \mathbf{\nabla} T^{-1}$). We get:

$$\frac{\partial_X P}{\partial t} = \int \mathbf{W} . \frac{\partial}{\partial t} \mathbf{\nabla} . T^{-1} d V = \int \mathbf{W} . \mathbf{\nabla} \frac{\partial T^{-1}}{\partial t} d V \qquad (7.8)$$

After an integration by parts, and cancellation of the boundary term due to the steady state assumption on the boundary Ω it becomes:

$$\frac{\partial_X P}{\partial t} = - \int \frac{\partial T^{-1}}{\partial t} \mathbf{\nabla} \mathbf{W} d V \, ; \qquad \left(\frac{\partial T^{-1}}{\partial t} \right)_\Omega = 0 \qquad (7.9)$$

Then, using the energy balance equation:

$$\rho \, c_v \, \frac{\partial T}{\partial t} + \nabla \, \mathbf{W} = 0 \tag{7.10}$$

and taking into account the stability condition for local equilibrium, ($c_v > 0$), we obtain as it was to be expected:

$$\frac{\partial_X P}{\partial t} = - \int \frac{\rho \, c_v}{T^2} \left(\frac{\partial T}{\partial t} \right)^2 d V \leqslant 0 \tag{7.11}$$

where the equality sign concerns the steady-state. The general demonstration may be derived likewise from the complete conservation equations (cf. [1]).

It must be kept in mind that the inequality (7.5) is quite independent from any phenomenological relation between flows and forces. This provides us with a property valid in the whole range of macroscopic irreversible processes, either in the linear or non-linear region.

Other expressions of this criterion are also available. For instance, as the flows related to the steady state satisfy the condition:

$$(d_X P)_{st} = \int \sum_\alpha J_{\alpha \, st} \, d_t X_\alpha \, d V = 0 \tag{7.12}$$

one obtains by substracting (7.12) from (7.3), the <u>excess form</u>, valid around a given steady-state as

$$\int \sum_\alpha \Delta J_\alpha \frac{\partial}{\partial t} (\Delta X_\alpha) \, d V \leqslant 0 \tag{7.13}$$

Under this form, the evolution criterion is often useful to investigate the behaviour of a small disturbance, as e.g. a normal mode, around the steady-state of reference (cf. [1]).

On the other hand, the evolution criterion (7.5) permits to approach the stability problem of non-equilibrium states by the same way as for the classical stability theory of equilibrium. Indeed, (7.5) enables us to express the stability condition by the requirement that all evolution starting from the reference state is inconsistent with the evolution criterion. Whence the stability condition for non varied boundary conditions takes the form:

$$\delta_X P = \int \sum_\alpha J_\alpha \, \delta X_\alpha \, dV \geqslant 0 \qquad (7.14)$$

in the general case, and more particularly:

$$\delta_X P = \int \sum_\alpha \delta J_\alpha \, \delta X_\alpha \, dV \geqslant 0 \qquad (7.15)$$

in the case of small perturbations around a steady-state (see e.g. (7.12)).

We thus recover the excess entropy production, together with the related stability condition for small disturbances, derived formerly from the Liapounoff stability theory.

As a matter of fact, the evolution criterion (7.5) and the second law of thermodynamics, namely:

$$P = d_i S \geqslant 0 \qquad (7.16)$$

are of alike bearing. This justifies for (7.5) the denomination of Universal evolution criterion. Indeed, even for processes involving convection effects, this criterion preserves its bilinear form. This is easily obtained by adding the equality:

$$\int \left[\rho h \mathbf{v} . \nabla T^{-1} + T^{-1} \mathbf{v} . \nabla p - \sum_\gamma \rho_\gamma \mathbf{v} . \nabla (\mu_\gamma T^{-1}) \right] dV = 0 \tag{7.17}$$

derived from the Gibbs-Duhem relation, ($\mathbf{v}$ = barycentric velocity) to the entropy production given by (7.1). We obtain in this way the associated bilinear expressions:

$$P = \int \sum_\alpha J'_\alpha X'_\alpha \, dV \geqslant 0 \tag{7.18}$$

for the entropy production, and

$$d_{X'} P = \int \sum_\alpha J'_\alpha \, d_t X'_\alpha \, dV \leqslant 0 \qquad (d_t \equiv \partial / \partial t) \tag{7.19}$$

for the corresponding evolution criterion under stationary boundary conditions.

Let us recall that the flows J'_α here considered, involve a conduction flow as well as a convection flow (for more details and some restrictions, see [1]).

8. The Local Potential.

As a rule, to approach the evolution problem by attempting to solve the differential equations as they arise from the conservation and phenomenological laws, together with the constraints prescribed to the system (initial and boundary conditions) is a question attended by great difficulties.

In many cases, only approximate methods of numerical analysis are available. In this respect, the variational techniques present a particular interest whenever the solution of the problem may be derived as the minimum of some potential associated to the basic equations.

Unfortunately, in the general case, the evolution criterion $d_x P \leqslant 0$, occurs in the form of a non exact differential whereas its reduction to an exact differential of a potential is only possible for particular situations. As an example, most of the non-linear problems are devoid of such an associated potential, and therefore cannot be solved by the classical variational techniques as the Rayleigh-Ritz method.

Nevertheless, as pointed out hereafter, it remains still possible to split the differential expression of the evolution criterion into two parts in such a way, that one of these corresponds to the exact differential of a potential Φ (say).

Then, this potential occurs as a functional characterized by the following properties:

(i) - The Euler-Lagrange equations derived from Φ, together with additional subsidiary conditions stated below, restore the basic differential equations.

(ii) - For fixed boundary conditions, all arbitrary increment $\Delta\Phi$ starting from this stationary solution, is a positive quantity (absolute minimum).

(iii) - The functional Φ , together with the subsidiary conditions, involves additional parameters connected to the local state of the system; whence, the denomination of local potential for Φ .

Steady-states, as well as time dependent problems may be approached by the local potential method. To be specific, we shall restrict ourselves to the statement of the local potential method in the case of the steady-state non-linear heat conduction problem, as it occurs for a temperature dependent thermal conductivity $\lambda(T)$, in an isotropic medium. Indeed, this example enables us to underline easily the typical characteristics of the method, the generalization to other non linear problems being then straightforward [1] .

On the other hand, the reader will also observe that different local potential may be constructed for the same problem. As a rule they differ from one another, either by an additional flow term or by a positive weighting function as in the following heat conduction problem (cf. after eq.(8.8)).

The energy balance equation assumes here the simple form:

(8.1) $$\rho \, \partial_t e = -\nabla \mathbf{W}$$

Let us multiply the two sides by the increment δT and integrate over the volume. For fixed boundary conditions, namely:

$$(\delta T)_\Omega = 0 \qquad (8.2)$$

we get

$$\int \rho\, \delta T\, \partial_t e\, dV = \int \mathbf{W}\, \delta \nabla T\, dV \qquad (\nabla \delta = \delta \nabla) \qquad (8.3)$$

Then, using the Fourier's law:

$$\mathbf{W} = -\lambda \nabla T \quad ; \quad \lambda = \lambda(T) \qquad (8.4)$$

we obtain:

$$-\int \rho\, \delta T\, \partial_t e\, dV = \frac{1}{2} \int \lambda\, \delta (\nabla T)^2 dV \qquad (8.5)$$

Let us now restrict ourselves to the steady-state problem. Then we have:

$$\partial_t e = \partial_t \delta e = c_v\, \partial \delta T \qquad (8.6)$$

Neglecting also higher order terms, we adopt for the conductivity, the equality:

$$\lambda = \lambda_0 + \delta \lambda \quad ; \quad \lambda_0 \equiv \lambda(T_0) \qquad (8.7)$$

where T_0 denotes the steady-state temperature. Then (8.5) finally becomes:

$$- \int \rho \, \delta T \, \partial_t (\delta e) \, dV = \frac{1}{2} \int \lambda_0 \delta (\nabla T)^2 dV + \frac{1}{2} \int \delta \lambda . \delta (\nabla T)^2 dV \tag{8.8}$$

Apart from the weighting function T^{-2}, the l.h.s. corresponds to the excess entropy production, i.e. $\partial_t \delta^2 S/2$. As the stability condition is always secured in the present simple heat conduction problem, the l.h.s. of (8.8) is a positive quantity. On the contrary, the sign of each term in the r.h.s. is by no means, determined by this stability condition.

Indeed, around the steady state, not only the second, but also the first term is a second order quantity, as it is easily seen, taking account of $\nabla W_0 = 0$. Therefore the second term cannot be neglected, and the sign of the first has to be investigated separately. Let us introduce the functional

$$\Phi (T, T_0) = \int \mathcal{L} (T, T_0) \, dV \tag{8.9}$$

with the Lagrangian:

$$\mathcal{L} (T, T_0) = \frac{1}{2} \lambda_0 (\nabla T)^2 \tag{8.10}$$

Therefore the first term in the r.h.s. of (8.8) is equal to $\delta \Phi$ where Φ is a functional of the two variables:

T_0 , which denotes the non-varied presumed steady-state solution actually unknown,

T , which is the varied function, occurring as a fluctuating temperature whose average is T_0 .

The condition under which the functional Φ is stationary with respect to the variations of T is given by the Euler-Lagrange equation:

$$\frac{\delta \mathcal{L}}{\delta T} = - \nabla (\lambda_0 \nabla T) = 0 \tag{8.11}$$

It remains to prescribe that the solution $T^+(x_i)$, of this equation has to coïncide with the presumed value T_0 . This provides us with the so-called <u>subsidiary condition</u>:

$$T^+ = T_0 \tag{8.12}$$

Then the extremal (8.11) takes the form:

$$\left(\frac{\delta \mathcal{L}}{\delta T}\right)_{T_0} = - \nabla(\lambda_0 \nabla T_0) = 0 \tag{8.13}$$

which,as expected,restores the steady-state equation of the heat conduction problem.

To investigate the nature of this extremum, let us calculate $\Delta\Phi$ around the steady-state, namely:

$$\Delta\Phi = \Phi(T, T_0) - \Phi(T_0, T_0)$$

$$= \frac{1}{2}\int \lambda_0 \left[(\nabla T)^2 - (\nabla T_0)^2\right] dV \tag{8.14}$$

Putting:

$$\vartheta \equiv T - T_0 \tag{8.15}$$

we may expand the brackets in the r.h.s. of (8.14) in terms of the deviation ϑ .

After integration by parts and cancellation of the boundary term, the linear part in ϑ vanishes due to (8.13). Hence

$$\Delta\Phi = \frac{1}{2}\int \lambda_0 (\nabla\vartheta)^2 dV > 0 \tag{8.16}$$

around the steady state and the extremum is an absolute minimum, whatever the amplitude of the deviation ϑ . This is an important property for the approximate methods of numerical calculus. On the other hand, according to (8.13) and (8.16), the functional $\Phi(T, T_0)$ occurs as a local potential, satisfying the definition given previously.

The solution T^+ of eq. (8.11) may be interpreted as a functional of the presumed steady-state distribution T_0 . In other words, the subsidiary condition (8.12) may be rewritten more explicitly as:

$$T^{+}\left(\{T_0\}\right) = T_0 \tag{8.17}$$

which gives rise to a physical interpretation of the local potential method. Indeed, T^{+} being the solution of the variational problem, corresponds to a vanishing value of the deviation ϑ and therefore also of the corresponding positive quantity [1]:

$$-\delta^2 S = \int \frac{c_v}{T^2}\, \vartheta^2\, dV \tag{8.18}$$

There exists a simple relation between the probability of a small fluctuation and $\delta^2 S$. This relation is a generalization to non-equilibrium states of the Einstein's formula:

$$P_r \sim \exp \frac{\Delta S}{k} \tag{8.19}$$

valid for fluctuations around <u>equilibrium</u>; k is the Boltzmann constant. For small fluctuations one may write:

$$\Delta S = (\delta S)_{eq} + \frac{1}{2}(\delta^2 S)_{eq} \tag{8.20}$$

Besides, for an isolated system $(\delta S)_{eq} = 0$, and (8.19) becomes:

$$P_r \sim \exp \frac{\delta^2 S}{2k} \tag{8.21}$$

$\delta^2 S$, being a negative definite quadratic form, suggests to adopt (8.21) as the suitable expression of the Einstein's formula for small fluctuations around non-equilibrium states. This point of view has been entirely justified by the recent investigations of Nicolis and Babloyantz based on a stochastic analysis ([1], 1969; [1], 1971).

The most probable state corresponds to $\delta^2 S = 0$, and thus T^+ is the most probable temperature distribution among the different possibilities generated by small fluctuations. On the other hand, T_0 is the macroscopic solution, i.e. the average taken over arbitrary fluctuations. As a result, the subsidiary condition (8.17) expresses that the most probable temperature distribution with respect to small fluctuations, has to coincide with the average temperature distribution, with respect to all fluctuations. The general case may be treated likewise [1] .

Let us now use this local potential mehtod, as a numerical technique of successive approximations.

9. The Local Potential and the Iteration Method.

For a temperature dependent thermal conductivity $\lambda(T)$, in an isotropic medium, the steady state heat equation takes the non-linear form (8.13):

$$\nabla(\lambda \nabla T) = 0 \qquad (9.1)$$

As a rule, only approximate solutions are to be

expected, using e.g. the variational properties of the local potential, together with methods of numerical analysis. In this respect we have discussed previously in detail the <u>self-consistent</u> method and established sufficient convergence criterions, for the successive approximations [1]. Likewise, we describe hereafter the <u>iteration</u> method by trial and errors. As we shall see it, we obtain by this way the same type of convergence criterions.

Let us choose the local potential (8.9), (8.10):

$$\Phi(T, T_0) = \frac{1}{2}\int \lambda_0 (\nabla T)^2 dV \qquad (9.2)$$

$$\lambda_0 \equiv \lambda(T_0)$$

and minimize this functional with respect to T, keeping the unknown solution T_0 of eq. (9.1) as a non varied quantity. Bringing then the subsididiary condition $T = T_0$, into the Euler-Lagrange equation, we recover the Fourier equation (9.1) as the extremal of this peculiar variational problem (see section 8).

This property enables us to calculate by iteration, successive approximations of the exact solution T_0 of eq. (9.1) and to prove the convergence of the method.

To begin with, we adopt for T_0 as a first test function, an arbitrary steady-state distribution $T_d(x_\gamma)$, which does not necessarily satisfy the boundary conditions. Then we take for T the n parameters family of admissible functions:

$$(9.3) \qquad T_n = \Sigma\, \alpha_k\, \varphi_k(x_j) \qquad (k = 1, \ldots, n)$$

each φ_k satisfying the boundary conditions. By convenience, these conditions may be always presented under the form:

$$(9.4) \qquad \left[\varphi_k(x_j)\right]_{\Omega} = 0$$

on the surface Ω of the system.

The corresponding approximate value of the local potential (9.2) is

$$(9.5) \qquad \Phi(T_n, T_a) = \frac{1}{2} \int \lambda_a (\nabla T_n)^2 dV$$

Minimization with respect to the α_k followed by an integration by parts and cancellation of the boundary term due to (9.4) yields the set of n equations:

$$(9.6) \qquad \int \varphi_k \nabla (\lambda_a \nabla T_n)\, dV = 0$$

to determine the n values α_{k1} of the α_k, given by this first approximation. According to (9.3) the corresponding approximation for T is given by the equality:

$$(9.7) \qquad T_{n1} = \Sigma\, \alpha_{k1}\, \varphi_k(x_j)$$

To obtain the next approximation, we adopt this time for T_o the value T_{n1} and minimize the local potential:

$$\Phi(T_n, T_{n1}) = \frac{1}{2}\int \lambda_{n1} (\nabla T_n)^2 dV \tag{9.8}$$

which gives rise to n new values α_{k2} of the α_k parameters. The second approximation for T is then given by:

$$T_{n2} = \Sigma\ \alpha_{k2}\ \varphi_k(x_j) \tag{9.9}$$

Henceforth, the iteration process becomes straightforward, and after m successive iterations we get:

$$\int \varphi_k \nabla(\lambda_{n,m-1} \nabla T_n)\ dV = 0 \tag{9.10}$$

to determine the n values of α_{km}, and subsequently:

$$T_{n,m} = \Sigma\ \alpha_{km}\ \varphi_k(x_j) \tag{9.11}$$

Let us now investigate the convergence near the exact value T_0 of the successive approximations T_a, $T_{n,1}$, $T_{n,2}$, .. $T_{n,m}$, supplied by this iteration method.

To proceed, we first minimize the local potential $\Phi(T, T_0)$ where the admissible functions are those of the family (9.3). We such obtain the set of n equations:

$$\int \varphi_k \nabla(\lambda_0\ \nabla T_n)\ dV = 0 \tag{9.12}$$

to determinate the n values $\bar{\alpha}_k$ of the parameters, and therefore the corresponding:

$$\bar{T}_n = \Sigma\, \bar{\alpha}_k\, \varphi_k(x_j) \qquad (k = 1,..,n) \tag{9.13}$$

Let us now observe that the sequence of order given in (9.3), corresponds to a larger family of admissible functions than the sequence of order $n-1$. Indeed, the latter occurs as a particular case of the first, obtained for $\alpha_n = 0$. As a result, the minimum $\Phi(\bar{T}_n, T_0)$, corresponds to a lower value than $\Phi(T_{n-1}, T_0)$, and, exactly as in the classical Rayleigh-Ritz method, one derives the successive inequalities:

$$\Phi(\bar{T}_1, T_0) \geqslant \Phi(\bar{T}_2, T_0) \dots \geqslant \Phi(\bar{T}_n, T_0) \dots \geqslant \Phi(T_0, T_0) \tag{9.14}$$

Moreover, assuming that the exact solution $T_0(x_j)$ of the Fourier equation may be represented everywhere in V , as a complete set of linearly independent functions $\varphi_k(x_j)$ satisfying the boundary conditions, we may conclude that a value of n does exist, beyond which $\Phi(T_n, T_0)$ becomes arbitrarily close to the exact minimum $\Phi(T_0, T_0)$. This implies:

$$\varepsilon_n \equiv \bar{T}_n - T_0 \to 0 \quad \text{for} \quad n \to \infty \tag{9.15}$$

Now we derive from (9.10) the identities:

$$0 = \int \varphi_k \nabla (\lambda_{n,m-1} \nabla T_{n,m})\, dV = \int \varphi_k \nabla (\lambda_{m-1} \nabla T_m)\, dV \tag{9.16}$$

disregarding the subscript n in the r.h.s. to simplify the notation. On the other hand (9.12) gives us:

$$0 = \int \varphi_k \nabla (\lambda_o \nabla \bar{T}_n) \, dV \tag{9.17}$$

We substract (9.17) from (9.16) and multiply by $(\alpha_{km} - \bar{\alpha}_k)$. Then we sum over k, to introduce the deviation:

$$\vartheta_m = T_m - \bar{T}_n \;; \quad (\vartheta_m)_\Omega = 0 \tag{9.18}$$

We obtain in this way:

$$\int \vartheta_m \nabla \left[\lambda_{m-1} \nabla T_m - \lambda_o \nabla \bar{T}_n \right] dV = 0 \tag{9.19}$$

or, after an integration by parts and cancellation of the boundary term:

$$\int \lambda_{m-1} (\nabla \vartheta_m)^2 \, dV = - \int (\lambda_{m-1} - \lambda_o) \, \nabla (T_o + \varepsilon_n) . \nabla \vartheta_m \, dV \tag{9.20}$$

according to (9.18) and (9.15).

More particularly, for the first step of the iteration method $(m = 1)$, (9.20) takes the form:

$$\int \lambda_\alpha (\nabla \vartheta_1)^2 \, dV = - \int (\lambda_\alpha - \lambda_o) \, \nabla (T_o + \varepsilon_n) . \nabla \vartheta_1 \, dV \tag{9.21}$$

Indeed, in that case, the quantity λ_{m-1} has to be replaced by λ_α associated to the arbitrary value T_α , according to (9.6). By convenience, eq. (9.20) may be rewritten as:

$$\int \lambda_{m-1} (\nabla \vartheta_m)^2 \, dV = - \int \left[\frac{\lambda_{m-1} - \lambda_o}{T_{m-1} - T_o} (\nabla T_o) (\vartheta_{m-1} + \varepsilon_n) (\nabla \vartheta_m) + \right.$$

$$(9.22) \qquad + (\lambda_{m-1} - \lambda_0)(\nabla \varepsilon_n)(\nabla \vartheta_m)\Big] dV$$

Let us replace in the l.h.s. the positive quantity λ_{m-1} by the lower bound λ_{min} deduced from the law $\lambda(T)$, which has to be given here for all the numerical values of T between $-\infty$ and $+\infty$, and according to the experimental data in the region of physical interest (see [1]).

We also introduce in the r.h.s. an upper bound of the quantity:

$$\frac{\lambda_{m-1} - \lambda_0}{T_{m-1} - T_0} \nabla T_0$$

This upper bound is a positive quantity which may be estimated roughly as the product of the maximum slope of the curve $\lambda(T)$, by the highest absolute value presumed for ∇T_0. For practical purposes, this rough estimation is generally sufficient and may be written as:

$$|\Delta \lambda|_{max} / L$$

where L, denotes a characteristic length, to choose according to the condition (9.28) given below.

We thus derive from (9.22) the inequality:

$$\lambda_{min} \int (\nabla \vartheta_m)^2 dV < \frac{|\Delta\lambda|_{max}}{L} \int \left| \vartheta_{m-1} + \varepsilon_n \right| \left| \nabla \vartheta_m \right| dV$$

$$+ \left| \Delta\lambda \right|_{max} \int \left| \nabla \varepsilon_n \right| \cdot \left| \nabla \vartheta_m \right| dV \qquad (9.23)$$

Let us now observe that the Schwartz inequality:

$$(f, g)^2 \leqslant (f, f)(g, g) \quad ; \quad (f, g) = \int f g \, dV \quad (9.24)$$

permits to factorize the quantity:

$$\widetilde{\nabla\vartheta}_m = \left[\int (\nabla \vartheta_m)^2 dV \right]^{1/2} \qquad (9.25)$$

in the two sides of (9.23). We use here the more compact notation:

$$\tilde{Z} = \left[\int Z^2 dV \right]^{1/2} \qquad (9.26)$$

After this simplification, $\left| \nabla \vartheta_m \right|$ disappears from the r.h.s. of (9.23).

On the other hand, we may also apply to the l.h.s. of (9.23) the classical inequality:

$$\widetilde{\nabla\vartheta}_m > \frac{\pi}{L} \tilde{\vartheta}_m \qquad (9.27)$$

valid for $(\vartheta_m)_\Omega = 0$ (see [1]). Here L denotes a length satisfying the condition:

$$L^{-2} \leqslant a^{-2} + b^{-2} + c^{-2} \qquad (9.28)$$

where a, b, c, are the edges parallel to the axis x, y, z of a parallelepiped enclosing V.

Then the inequality (9.23) becomes:

$$\tilde{\vartheta}_m < \frac{|\Delta\lambda|_{max}}{\pi\lambda_{min}}\left(\tilde{\vartheta}_{m-1} + \tilde{\varepsilon}_n + L\,\widetilde{\nabla\varepsilon}_n\right) \tag{9.29}$$

and likewise, starting from (9.21):

$$\tilde{\vartheta}_1 < \frac{|\Delta\lambda|_{max}}{\pi\lambda_{min}}\left(\widetilde{T_a - T_o} + L\,\widetilde{\nabla\varepsilon}_n\right) \tag{9.30}$$

for the first iteration $(T_a - T_o = \vartheta_a + \varepsilon_a)$.

These m relations enable us to eliminate the $m-1$ quantities $\tilde{\vartheta}_{m-1}$ by recurrence.

Putting for the sake of simplicity:

$$B \equiv \frac{|\Delta\lambda|_{max}}{\pi\lambda_{min}} \tag{9.31}$$

we get finally:

$$\tilde{\vartheta}_m < \left(\widetilde{T_a - T_o}\right)B^m + \left(\sum_{\ell=1}^{m} B^{\ell}\right)\left(\tilde{\varepsilon}_n + L\,\widetilde{\nabla\varepsilon}_n\right) \tag{9.32}$$

Therefore the inequality:

$$B < 1 \qquad \text{or} \qquad \frac{\lambda_{max} - \lambda_{min}}{\lambda_{min}} < \pi \tag{9.33}$$

represents a sufficient condition for the convergence in the mean of iteration method. Indeed:

1°) For $m \to \infty$, $\sum_{\ell=1}^{m} B^{\ell}$ remains finite and $B^m \to 0$ if

(9.33) is fulfilled.

2°) For $n \to \infty$, $\tilde{\varepsilon}_n$ as well as $\nabla\varepsilon_n \to 0$, since eq. (9.15) is valid in each point of the system.

Finally, let us emphasize that the above discussion does not close the problem, because more refined treatments are possible. In this respect investigations based on the application of the functional analysis to the local potential properties, are actually undertaken successfully by Steve Rohde (*).

All we wanted to demonstrate here, is that a rough approximation adopted as upper and lower limit of the thermal conductivity λ , cover already a very large domain where the convergence of the iteration method is secured.

The influence arising from the choice of the first arbitrary test function T_a appears clearly in the r.h.s. of (9.32).

Anyway let us emphasize that the foregoing demonstration is entirely based on the positive character of the heat conductivity λ .

(*) General Motors, Research Laboratories, Warren, Michigan. Private communication.

10. Stability problems in fluids at rest – Bénard instability.

10.1. Bénard instability and entropy production.

Let us consider a horizontal layer of fluid in which a temperature gradient is maintained by heating from below (the so called "adverse" temperature gradient). For sufficiently small values of this slope, heat is transmitted only by conduction. However beyond a critical value convection appears.

The non dimensional parameter which governs the stability of the layer is the Rayleigh number (cf. [1]):

(10.1) $$\mathcal{R}_a = \frac{g\,\alpha\,\beta}{\chi\,\nu}\, d^4 > 0$$

where g is the acceleration due to gravity, d the depth of the layer, $\beta = |dT/dz|$ the adverse gradient, α, χ, ν respectively the coefficients of volume expansion, the thermal diffusivity ($\chi = \lambda/\rho c_v$) and $\nu = \eta/\rho$ the kinematic viscosity.

In order to investigate the stability we consider the generalized excess entropy production $P[\delta Z]$. Taking into account the incompressibility condition as well as the so called Boussinesq approximation (see [2]), we obtain for $P[\delta Z]$ (cf. [1] :

$$P[\delta Z] = \int \rho \left[(\mathcal{R}_a)^{-1} \frac{(g\,\alpha)^2 d^4}{\nu} (\vartheta_{,j})^2 + \nu\,(u_{i,j})^2 \right.$$

$$\left. - 2\, g\, \alpha\, \vartheta\, w \right] dV > 0$$

(10.2) $$(\vartheta = \delta T\,; \quad u_i = \delta v_i\,; \quad w = \delta v_z)$$

using as weighting functions:

$$\varepsilon^2 = \frac{\alpha g}{c_v \beta} T^2 \; ; \qquad \tau^2 = T$$

defined in section 8.

The r.h.s. may be interpreted as the increment of the function:

$$F = \int \rho \left[(\mathcal{R}_a)^{-1} \frac{(g\alpha)^2 d^4}{\nu} (T_{,j})^2 + \nu (v_{i,j})^2 - 2 g \alpha T v_z \right] dV \tag{10.3}$$

After averaging over the $x - y$ plane we may write (cf. [1] :

$$\langle F \rangle = \langle F_0 \rangle + \langle P[\delta Z] \rangle \tag{10.4}$$

where we have put:

$$\langle F_0 \rangle = \frac{\rho_1 \nu \varkappa^2}{d^3} \mathcal{R}_a$$

The stability condition (10.2) may then be interpreted as a minimum condition for the function $\langle F \rangle$ which may then be identified as a true potential. Instability occurs therefore when $\langle P[\delta Z] \rangle$ vanishes i.e. $\langle F \rangle = \langle F_0 \rangle$ (degeneracy of $\langle F \rangle$). The analogy with phase transition is evident (see fig. 1).

Let us stress that the potential $\langle F \rangle$ appears as a difference between two positive quantities, written symbolically as:

$$\langle F \rangle = \text{(dissipative effects)} - \text{(convective effects)}$$

10.2. The Bénard dissipative structure.

One could show that the convection-free state of reference belongs to the region of strictly linear thermodynamics and consequently the theorem of minimum entropy production is here valid (see section 7).

At the marginal state (Bénard point, neutral stability) the entropy production $P[S]$ varies suddenly with the appearance of the first unstable normal mode (see fig.2).

It is clear that for $\mathcal{R}_a \geqslant (\mathcal{R}_a)_c$ the description of the system in terms of linear thermodynamics breaks down and new coupling effects arise. In other words the thermodynamic branch related to the system at rest, is replaced by a new branch including convection. This transition brings about the so called dissipative structure. In the present case it corresponds to an exchange of stability characterized by a well known stationary cellular motion (see [1]). The potential $\langle F \rangle$ defined above plays then a role similar to that of the Helmholtz free energy, in phase transitions.

After this short summary, developped elsewhere (see [1]), let us investigate in detail the stability of the more elaborate two component Bénard problem. This application has been extensively treated from both theoretical and experimental points of view by J.C. Legros (see [3]).

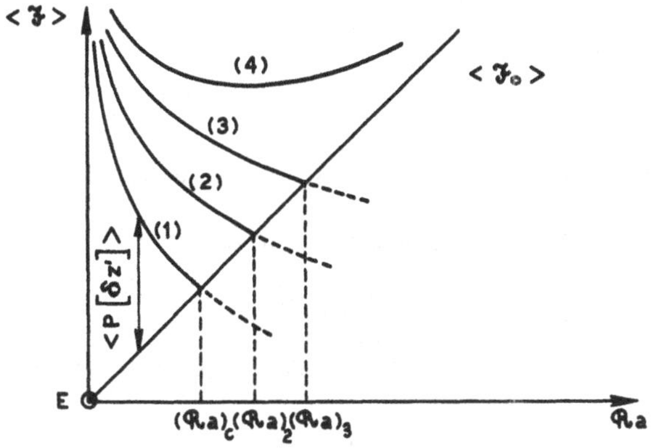

Fig. 1.

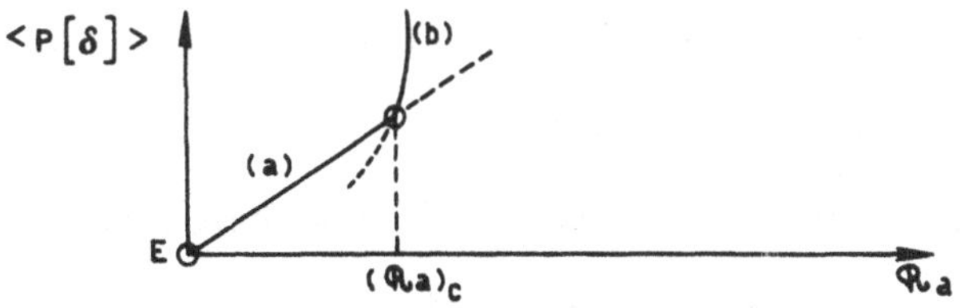

Fig. 2.
(a) Thermodynamic branch including the equilibrium state E.
(b) Branch generated by the first critical normal mode.

References

(Sections 1 to 10)

[1] P. Glansdorff and I. Prigogine, Thermodynamic Theory of Structure Stability and Fluctuations, Wiley, Interscience, London, 1971.

[2] Chandrasekhar, Hydrodynamic and Hydromagnetic Stability, Clarendon Press, Oxford, 1961.

[3] J.C. Legros, Thèse de doctorat, Chimie Physique, Université Libre de Bruxelles, 1971.

11. Stability of a two component fluid layer heated from below.

11.1. Introduction.

We shall now be concerned with the so-called Bénard problem in a two component fluid layer. This problem has been studied recently by J.C. Legros [1, 2, 3] from the experimental point of view and theoretically by J.C. Legros, J.K. Platten, and P. Poty [4] (see also [9]).

The kinetic theory of the Bénard problem in a one component fluid layer has been solved in details by Chandrasekhar [5, chap. II] using the usual normal mode analysis.

The problem is to find out when free convection is generated due to the instability of the state at rest. As quoted in Section 10, the corresponding transition point is characterized by the critical value of the dimensionless parameter i.e. the Rayleigh number:

$$\mathcal{R}_a = \frac{g \alpha \beta d^4}{\varkappa \nu} \tag{11.1}$$

together with the critical value of the wave number k included in the expression of the vertical velocity perturbation:

$$w = W(Z) \exp(ik_x x + ik_y y + \sigma t) \tag{11.2}$$

for a normal mode, with

$$k = \sqrt{k_x^2 + k_y^2}$$

Finally starting from the linearized perturbation equations, Chandrasekhar obtains the dispersion equation (for more details, see [5]):

$$(D^2 - k^2)(D^2 - k^2 - \sigma)(D^2 - k^2 - \sigma Pr)W = -\mathcal{R}_a k^2 W \qquad (11.3)$$

where $D^2 = d^2/dz^2$, k = the wave number, σ = the amplification coefficient, and $Pr = \nu/\chi$ the Prandtl number. The above equation gives rise for suitable boundary conditions to the critical values:

$$\mathcal{R}_a^{cr.} = 657.5 \quad \text{and} \quad k^{cr.} = \frac{\pi^2}{2} \text{ (free surfaces)} \qquad (11.4)$$

or

$$\mathcal{R}_a^{cr.} = 1707.762 \text{ and } k^{cr.} = 3.117 \text{ (rigid surfaces)} \qquad (11.5)$$

By comparison, using the local potential variational technique, Schechter and Himmelblau [6] found for two rigid surfaces and with only one variational parameter the excellent approximation

$$\mathcal{R}_a^{cr.} = 1750 \qquad \text{and} \quad k^{cr.} = 3.1$$

Accordingly, we intend to follow the same way, in order to study the similar problem of stability for a two component fluid layer. In this respect let us emphasize that, already for such an apparently slight generalization to a mixture, an exact solution is excluded and only approximate solutions are to be expected.

This remark enhances the interest of the local

potential method for a large class of problem in macroscopic physics.

11.2. The experimental problem.

11.2.1. Description of the apparatus.

As it can be seen on fig. I, the cell is constituted by two circular brass plates (L_1 and L_2 ; diameter = 20 cm; thickness = 20 mm) coated with gold to avoid corrosion. The plates are plane within a tolerance of 0.01 mm.

The L plates are fitted in thick stainless steel disks (A_1 and A_2) to ensure the cell a high mechanical rigidity.

The L_1 plate is provided with a heating coil of 300 Ω (in Kanthal type 0, wire diameter = 0.5 mm). The latter thermally and electrically insulated on the A_1 side by a disk of compressed asbestos (thickness 5mm) and the L_1 side, is electrically insulated by a thin plate of mica.

In L_2 is machined a rectangular channel (twin spirals) in such a way that in two neighbouring channels, water at constant temperature flows in opposite directions ensuring thus a good thermal homogeneity.

Tightness, distance and parallelism between the L plates are secured by means of a PTFE gasket (J). The parallelism between the plates is given within a tolerance of 0.03/200. The temperature difference is measured by means of two platinum resistors (Pt' and Pt'' ; Degussa, type P4 , $R = 100\,\Omega$ at 0° C) and the resistance differences are measured with a Leeds and Northrup Mueller bridge within an accuracy of $\pm$ 0.0001 Ω. The cell is thermally insulated by means of styrofoam S. M is a copper screen whose temperature is kept equal to that of L_1 and T is a stainless steel tube with thin walls to diminish the heat condition.

The horizontal level of the L_1 plate is very carefully ensured.

The entire apparatus is placed in a large air thermostat whose temperature is kept to that of L_1 .

11.2.2. Experimental results.

The following mixtures have been studied:

$$\begin{array}{c} C_6H_6 - CC\ell_4 \\ C_6H_5C\ell - CC\ell_4 \\ C_6H_{12} - CC\ell_4 \\ C_2H_2Br_4 - C_2H_2C\ell_4 \\ H_2O - CH_3OH \\ H_2O - C_2H_5OH \end{array} \qquad (11.6)$$

In Ref. 1, 2, 3, one finds the complete series of curves in which temperature differences have been plotted versus heat flux for the above systems at different concentrations.

In tables 1 to 6 one finds the values of the critical temperature difference for different initial concentrations of the systems under study for given thickness of the fluid layer.

For the four first systems D'/D is positive (*) and the system is destabilized by thermal diffusion. On the other hand, for the two last systems D'/D changes of sign at a concentration treshold, and becomes thus negative for large water concentration. In this case the system is stabilized by the thermal diffusion.

Fig. 2 represents ΔT versus W for the CCl_4-C_6H_6 system ($X_{CCl_4} = 0.45$). D'/D is here positive and a destabilizing effect was to be expected, but as it can be seen in Ref. 1, 2, 3 this effect was never observed in systems for which D'/D is positive.

Fig. 3 gives $\Delta T/W$ for the H_2O-CH_3OH system with

(*) D'/D is the Soret coefficient (D' is the thermal diffusion coefficient, D the isothermal diffusion coefficient). It is a measure of the separation which appears in a mixture placed in a temperature gradient. Supposing linear variation, the mass fraction difference is given by

$$\Delta N = \frac{D'}{D} N(1 - N) \Delta T$$

By definition, D'/D is positive when the denser component migrates towards the cold plate.

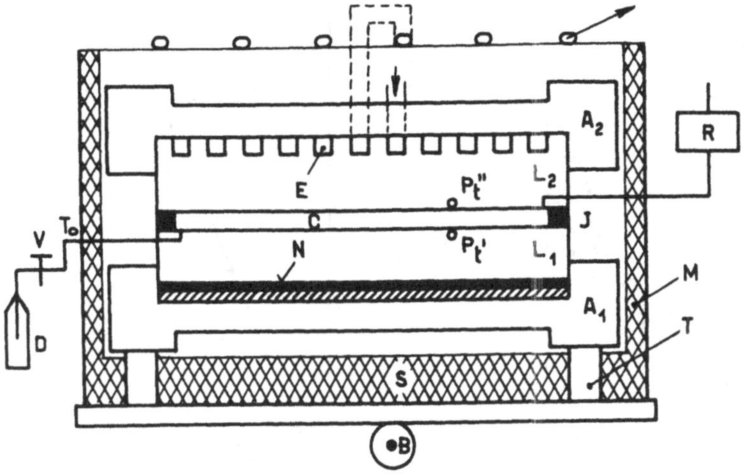

Fig. 1. The Bénard cell.

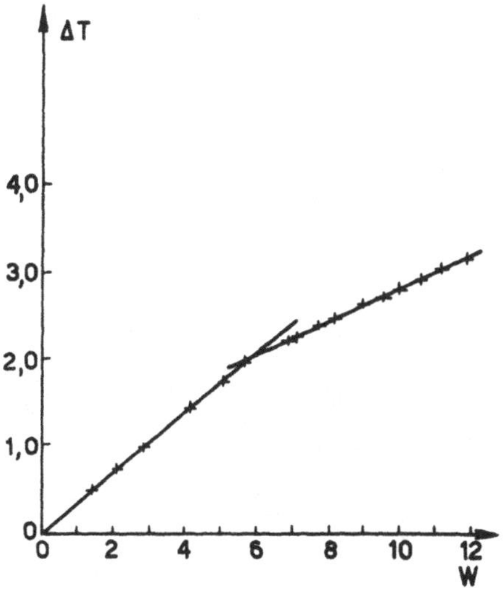

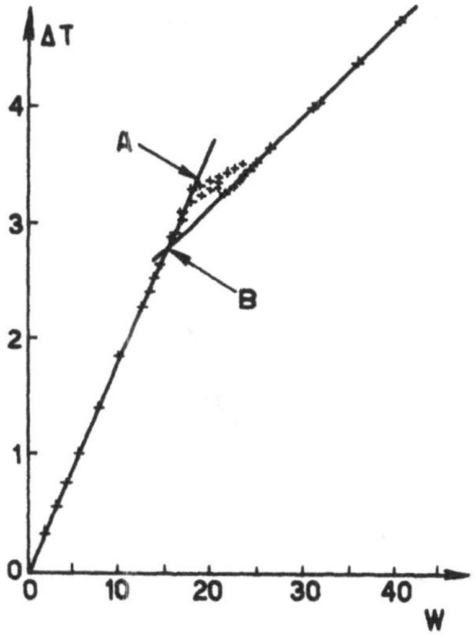

Fig. 2. System $CCl_4 - C_6H_6$ $X_{CCl_4} = 0,45$
Thickness of the layer = 0.9 mm
$\Delta T^{cr.} = 2.06°$ C.

Fig. 3. System $H_2O - CH_3OH$ $X_{H_2O} = 0,90$
Thickness of the layer = 3.13 mm ± 0.03 mm
A = 3.20° C
B = 2.81° C
ΔT versus the heat flux W.

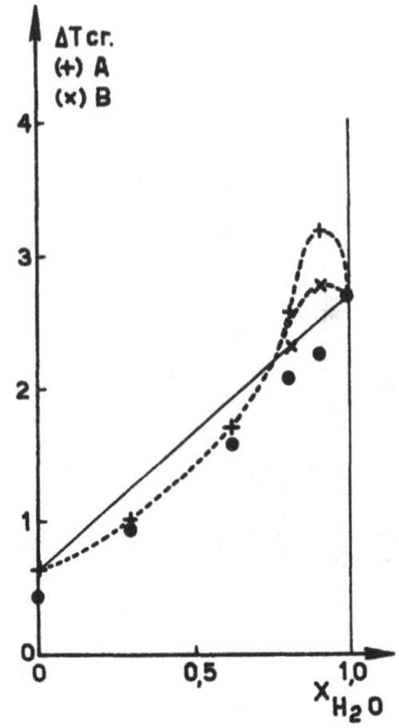

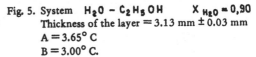

Fig. 4. System $H_2O - CH_3OH$
Thickness of the layer = 3.13 mm ± 0.03 mm
$\Delta T^{cr.}$ versus the mass fraction.

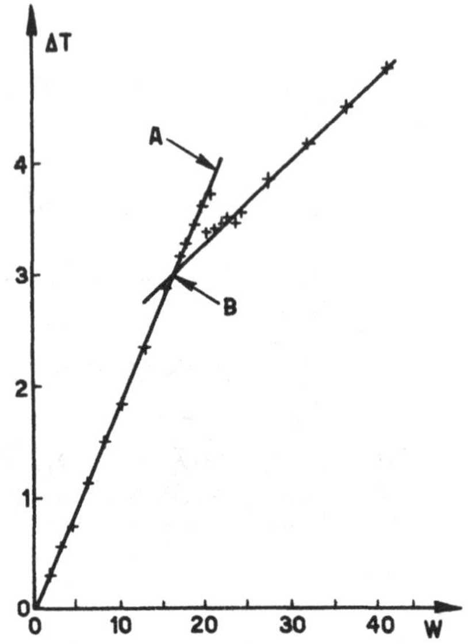

Fig. 5. System $H_2O - C_2H_5OH$ $X_{H_2O} = 0{,}90$
Thickness of the layer = 3.13 mm ± 0.03 mm
A = 3.65° C
B = 3.00° C.

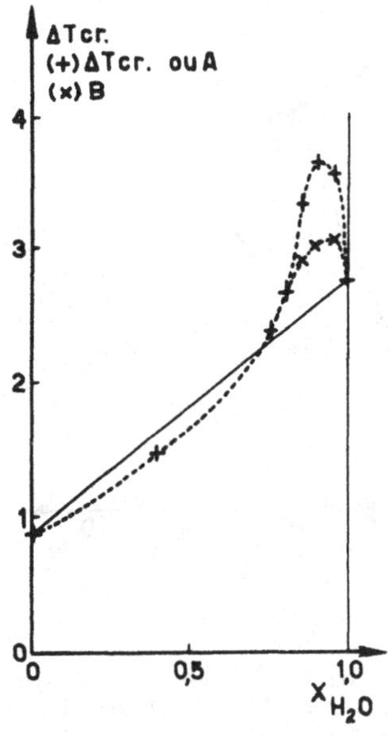

Fig. 6. System $H_2O - C_2H_5OH$
Thickness of the layer = 3.13 mm ± 0.03 mm
$\Delta T^{cr.}$ versus the mass fraction.

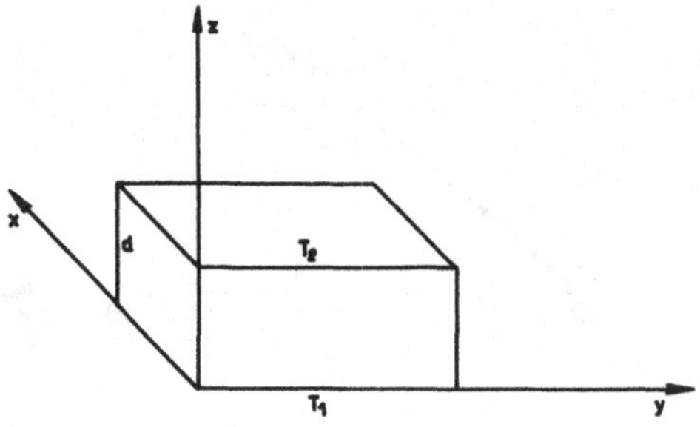

Fig. 7. System of coordinates used.

a large water concentration. The stabilizing effect by thermal diffusion is here observed. It must be noticed that there exists a hysteresis effect: ΔT is not the same for a given heat flux W everywhere. It depends whether we flow the curve for increasing or decreasing W. This same effect has also been observed in Bénard experiments in the $H_2O - C_3H_7OH$ system by Platten [16] and in the sea-water system by Caldwell [17].

In fig. 4 are plotted critical temperature differences versus initial concentration. The stabilizing effect is observed in the range of concentration where D'/D is negative. Moreover, points (.) are calculated using formula (11.1) together with $\mathcal{R}_a^{cr} = 1708$, and disregarding the thermal diffusion.

Likewise, fig. 5 and 6 represent the behaviour of the $H_2O - C_2H_5OH$ system.

11.3. The local potential technique applied to the problem of the stability of a two component fluid layer, taking into account the thermal diffusion contribution

11.3.1. Introduction

R.S. Schechter, I. Prigogine and J. Hamm, [9] examined theoretically the transition to free convection in a two component mixture. Initially, they assumed that the principle of exchange of stability remains valid and in order to find an exact solution, in that case, made some additional assumptions in the starting equations: in the thermal diffusion term (see eq. (11.8) the product $N_1 (1-N_1)$ is kept constant and in writing the linearized eq. 11.18 fluctuations of the mass fraction in this product are forbidden.

In this third section, we shall examine this question by means of a variational method based on the notion of local potential introduced by P. Glansdorff and I. Prigogine [10, 11, 12] but removing the additional assumptions made by R.S. Schechter, I. Prigogine and J. Hamm.

11.3.2. The steady state.

The conservation laws for mass momentum and energy may be written for a two component incompressible system as:

$$0 = \frac{\partial V_j}{\partial x_j} \tag{11.7}$$

$$\frac{\partial N_1}{\partial t} = - V_j \frac{\partial N_1}{\partial x_j} + D \frac{\partial^2 N_1}{\partial x_j^2} + D' \frac{\partial}{\partial x_j} \left[N_1 (1 - N_1) \frac{\partial T}{\partial x_j} \right] \tag{11.8}$$

$$\frac{\partial V_i}{\partial t} = - V_j \frac{\partial V_i}{\partial x_j} - \frac{1}{\rho_0} \frac{\partial p}{\partial x_i} + \frac{\rho}{\rho_0} F_i + \nu \frac{\partial^2 V_i}{\partial x_j^2} \tag{11.9}$$

$$\frac{\partial T}{\partial t} = - V_j \frac{\partial T}{\partial x_j} + \varkappa \frac{\partial^2 T}{\partial x_j^2} \tag{11.10}$$

Index 1 refers to the more dense component.

In order to obtain the system of equations (11.7 - 11.10) the usual assumptions are made:

1) we consider the density as a constant ρ_0 taken at a reference temperature and at the initial composition of the system (i.e. before thermal diffusion sets in), except in those terms containing the external forces acting on the system. This is the so-called Boussinesq approximation [5] .

2) in the energy equation, we neglect the Dufour effect.

Moreover, we restrict ourselves to dilute solutions $(N \ll 1)$, but the generalization to concentrated solutions is straightforward.

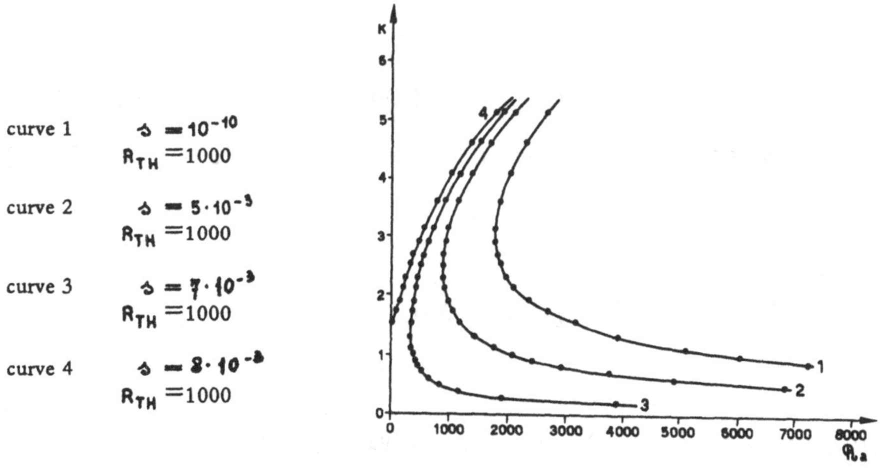

Fig. 8. Marginal curves of stability for different values of R_{TH} and ϑ, in the k/Ra plane.

At the steady state (index s), system (7.10) can be satisfied by:

a) (11.11) $$V_i^s = 0$$

b) (11.12) $$T^s = T_1 - \beta \lambda_j x_j = T_1 - \beta z$$

with $$\lambda_j = (\lambda_z, \lambda_x, \lambda_y) = (1,0,0)$$

and $$\beta = \frac{T_1 - T_2}{d} > 0$$

c) (11.13) $$\frac{\partial p^s}{\partial x_i} = \rho F_i = - g \rho(T, N) \lambda_i$$

d) (11.14) $$D \frac{\partial^2 N_1^s}{\partial z^2} + D' \frac{\partial}{\partial z}\left(N_1^s \frac{\partial T^s}{\partial z}\right) = 0$$

Thanks to the boundary conditions (the thermal diffusion flux is equal to zero at the two boundaries) and eq. 11.12, eq. 11.14 gives

(11.15) $$D \frac{\partial N_1^s}{\partial z} - D' N_1^s \beta = 0$$

The integration of eq. 11.15 over z yields an exponential mass fraction distribution (Hiby and Wirtz solution, see De Groot [13]).

However, for the usual values of D'/D and β this exponential mass fraction distribution may be approximated with a good accuracy by a linear law.

So

$$N_1^s = N_1^{in.}\left[1 + \frac{D'}{D}\left(z - \frac{d}{2}\right)\beta\right] \tag{11.16}$$

where $N_1^{in.}$ is the initial mass fraction of the more dense component, i.e., before thermal diffusion sets in.

Thus, the separation is given by

$$\Delta N = N_1^{in}\frac{D'}{D}\beta d$$

11.3.3 The Linearized Equations

We consider small perturbations around the steady state defined in the preceding section and, as usual in linear stability analysis, we restrict ourselves to terms of the first order in the perturbations.

Then

$$\begin{aligned}
N_1 &= N_1^s + \varepsilon\, n_1 + \dots \\
\rho &= \rho^s + \varepsilon\, \delta\rho + \dots \\
T &= T^s + \varepsilon\, \vartheta + \dots \\
V_j &= \phantom{T^s +{}} \varepsilon\, u_j' + \dots \\
p &= p^s + \varepsilon\, p' + \dots
\end{aligned} \tag{11.17}$$

The x, y, z components of u_j' are denoted respectively u', v', w'. At the first order in ε, in the geometry of Fig. 7, system (11.7) - (11.10) becomes

$$\frac{\partial n_1}{\partial t} = - w' \frac{\partial N_1^0}{\partial z} + D \nabla^2 n_1 - D' \frac{\partial n_1}{\partial z} \beta$$

(11.18)
$$+ D' \frac{\partial N_1^0}{\partial z} \cdot \frac{\partial \vartheta}{\partial z} + D' N_1^0 \nabla^2 \vartheta$$

(11.19)
$$\frac{\partial w'}{\partial t} = - \frac{1}{\rho_0} \frac{\partial p'}{\partial z} + \frac{\delta \rho}{\rho_0} (- g) + \nu \nabla^2 w'$$

(11.20)
$$\frac{\partial u}{\partial t} = - \frac{1}{\rho_0} \frac{\partial p'}{\partial x} + \nu \nabla^2 u'$$

(11.21)
$$\frac{\partial v'}{\partial t} = - \frac{1}{\rho_0} \frac{\partial p'}{\partial y} + \nu \nabla^2 v'$$

(11.22)
$$\frac{\partial \vartheta}{\partial t} = \beta w' + \chi \nabla^2 \vartheta$$

We also have the continuity equation

(11.23)
$$0 = \frac{\partial u_j'}{\partial x_j}$$

We now introduce a linear total steady density

$$\rho^s = \rho_0 (1 - \alpha \Delta T^s + \gamma \Delta N_1^s) \tag{11.24}$$

with

$$\alpha = - \frac{1}{\rho_0} \frac{\partial \rho}{\partial T} > 0 \tag{11.25}$$

and

$$\gamma = \frac{1}{\rho_0} \frac{\partial \rho}{\partial N_1} > 0 \tag{11.26}$$

Thus

$$\delta \rho = \rho_0 (- \alpha \vartheta + \gamma n_1) \tag{11.27}$$

This last equation is introduced in eq. (11.19). In order to have dimensionless quantities, we reduce the time by a factor d^2/ν, the velocity disturbance by $\varkappa / d$, the temperature disturbance by βd, the steady state mass fraction distribution, and the mass fraction disturbance by N_1^{in}, the pressure disturbance by $\rho_0 \varkappa \nu / d^2$, and finally the coordinates of the system by a characteristic length d.

System (11.18 - 11.23) becomes: (also using (11.16) for the steady state mass fraction distribution)

$$S_c \frac{\partial n_1}{\partial t} = - \frac{S_c}{P_r} w' s + \nabla^2 n_1 - s \frac{\partial n_1}{\partial z} + s^2 \frac{\partial \vartheta}{\partial z} + s N_1^s \nabla^2 \vartheta \tag{11.28}$$

$$\frac{\partial w'}{\partial t} = -\frac{\partial p'}{\partial z} + Ra\,\vartheta - R_{TH}\,n_1 + \nabla^2 w' \tag{11.29}$$

$$\frac{\partial u'}{\partial t} = -\frac{\partial p'}{\partial x} + \nabla^2 u' \tag{11.30}$$

$$\frac{\partial v'}{\partial t} = -\frac{\partial p'}{\partial y} + \nabla^2 v' \tag{11.31}$$

$$P_r \frac{\partial \vartheta}{\partial t} = w' + \nabla^2 \vartheta \tag{11.32}$$

In 11.28 – 11.32 all the variables are dimensionless quantities. So quite naturally we introduce dimensionless numbers as the

(11.33) Prandtl number $P_r = \dfrac{\nu}{\varkappa}$

(11.34) Schmidt number $S_c = \dfrac{\nu}{D}$

(11.35) Rayleigh number $R_a = \dfrac{g\alpha\beta d^4}{\varkappa\nu}$

Also we have

$$R_{TH} = \frac{g\gamma N_1^{in} d^3}{\varkappa\nu} \tag{11.36}$$

$$s = \frac{D'}{D} \beta d \qquad (11.37)$$

that we call respectively the "Thermal diffusion Rayleigh number" and "the Soret number".

For the disturbances we now take (as usual in normal modes analysis, see i.e. Ref. 12c)

$$w' = W(z) e^{ik_x x + ik_y y} e^{-\sigma t} \qquad (11.38)$$

$$v' = \frac{ik_y}{k^2} \frac{dW}{dz} e^{ik_x x + ik_y y} e^{-\sigma t} \qquad (11.39)$$

$$u' = \frac{ik_x}{k} \frac{dW}{dz} e^{ik_x x + ik_y y} e^{-\sigma t} \qquad (11.40)$$

$$p' = \nabla(z) e^{ik_x x + ik_y y} e^{-\sigma t} \qquad (11.41)$$

$$\vartheta = \Theta(z) e^{ik_x x + ik_y y} e^{-\sigma t} \qquad (11.42)$$

$$n_1 = X(z) e^{ik_x x + ik_y y} e^{-\sigma t} \qquad (11.43)$$

with

$$k_x^2 + k_y^2 = k^2 \qquad (11.44)$$

The structure of u', v' and w' is such that the continuity equation 11.43 is automatically satisfied. Substitution of eq. 11.38-11.44 into the system 11.28 - 11.32 yields, after elimination of the pressure disturbance amplitude Δ.

$$(11.45) \quad (D^2 - k^2 + \sigma S_c - s D) X + s\left[N_1^s (D^2 - k^2) + s D\right] \vartheta = \frac{S_c}{P_r} s W$$

$$(11.46) \quad (D^2 - k^2)(D^2 - k^2 + \sigma) W = R_a k^2 \Theta - R_{TH} k^2 X$$

$$(11.47) \quad (D^2 - k^2 + \sigma P_r) \Theta = -W$$

where

$$D^n = \frac{d^n}{dz^n}$$

Thus, we have a system of coupled linear homogeneous differential equations, which, together with appropriate boundary conditions, defines an eigenvalue problem. For given R_{TH}, Ra, Sc, Pr, s, and k nontrivial solutions exist only for particular values of σ. The sign of the real part of $\sigma (\sigma_R)$ defines the stable and the unstable domain in the $k - Ra$ plane for given values of the other parameters. If $\sigma_R > 0$, the perturbations are damped, however if $\sigma_R < 0$ the perturbations are amplified and lead the system to a new state, stationary or not.

System (11.45 - 11.47) has such a structure that it requires a special development. In contradiction to the

Bénard problem in pure liquids, (11.45 - 11.47) is not self-adjoint. It contains the steady state mass fraction distribution which is a function of the z coordinate. Moreover, system (11.45 - 11.47) cannot be satisfied either by even solutions, or by odd solutions. The symmetry properties of the eigenfunction are destroyed.

There is no hope to find exact solutions for (11.45 - 11.47) either for rigid boundaries, or for free ones. However, it is possible by introducing additional assumptions (more or less realistic) to obtain a self-adjoint eigenvalue problem, whose exact solution can be found in Ref. 9. However, it is more suitable to approach the stability analysis with the exact differential equations (11.45 - 11.47) and if necessary to look only for approximate solutions, by a variational technique as provided by the local potential theory.

11.3.4 The Local Potential

The variational method has been developped by P. Glansdorff and I. Prigogine [10, 11, 12] and widely used in stability analysis [6, 14, 15] In this section, we shall give a minmum of detailed calculations and the reader should refer to the above mentioned papers for a detailed review of this variational technique. We start with the linearized equations (11.18 - 11.23), and we multiply them respectively by increments $-\delta n_1$, $-\delta w'$, $-\delta u'$, $-\delta v'$, $-\delta \vartheta$ and we add.

We use relations such as

$$- \frac{\partial n_1}{\partial t} \delta n_1 = - \frac{1}{2} \frac{\partial}{\partial t} (\delta n_1)^2 - \frac{\partial n_1^{(o)}}{\partial t} \delta n_1 \tag{11.48}$$

with

$$n_1 = n_1^0 + \delta n_1$$

We then integrate over the volume and by parts with respect to z only, using the following boundary conditions for $z = 0$ and $z = 1$.

$$\vartheta = 0 \tag{11.49}$$

$$w' = u' = v' = 0 \tag{11.50}$$

$$\frac{\partial n_1}{\partial z} - s n_1 + s N_1^s \frac{\partial \vartheta}{\partial z} = 0 \tag{11.51}$$

Eq. (11.51) arises from the zero thermal diffusion flux condition at the boundaries. It is then easy to prove that the local potential has the form

$$\Phi = \int dv \left[Sc \frac{\partial n_1^{(o)}}{\partial t} n_1 + w'^{(o)} \frac{Sc}{Pr} s n_1 + \right.$$

$$- \frac{\partial^2 n_1^{(0)}}{\partial x^2} n_1 - \frac{\partial^2 n_1^{(0)}}{\partial y^2} n_1 + \frac{1}{2}\left(\frac{\partial n_1}{\partial z}\right)^2$$

$$- s\, n_1^{(0)} \frac{\partial n_1}{\partial z} + s^2 \vartheta^{(0)} \frac{\partial n_1}{\partial z} - s N_1^{s} \frac{\partial^2 \vartheta^{(0)}}{\partial x^2} n_1 - s N_1^{s} \frac{\partial^2 \vartheta^{(0)}}{\partial y^2} n_1$$

$$+ s^2 \frac{\partial \vartheta^{(0)}}{\partial z} n_1 + s N_1^{s} \frac{\partial \vartheta^{(0)}}{\partial z} \frac{\partial n_1}{\partial z} + P_r \frac{\partial \vartheta^{(0)}}{\partial t} \vartheta - w'^{(0)} \vartheta$$

$$- \frac{\partial^2 \vartheta^{(0)}}{\partial x^2} \vartheta - \frac{\partial^2 \vartheta^{(0)}}{\partial y^2} \vartheta + \frac{1}{2}\left(\frac{\partial \vartheta}{\partial z}\right)^2 + \frac{\partial u'^{(0)}}{\partial t} u' + \frac{\partial p'^{(0)}}{\partial x} u' - \frac{\partial^2 u'^{(0)}}{\partial x^2} u'$$

$$- \frac{\partial^2 u'^{(0)}}{\partial y^2} u' + \frac{1}{2}\left(\frac{\partial u'}{\partial z}\right)^2 + \frac{\partial v'^{(0)}}{\partial t} v' + \frac{\partial p'^{(0)}}{\partial y} v' - \frac{\partial^2 v'^{(0)}}{\partial x^2} v'$$

$$- \frac{\partial^2 v'^{(0)}}{\partial y^2} v' + \frac{1}{2}\left(\frac{\partial v'}{\partial z}\right)^2 + \frac{\partial w'^{(0)}}{\partial t} w' + p'^{(0)} \frac{\partial w'}{\partial z} - \mathcal{R}_a \vartheta^{(0)} w'$$

$$+ R_{TH} n_1^{(0)} w' - \frac{\partial^2 w'^{(0)}}{\partial x^2} w' - \frac{\partial^2 w'^{(0)}}{\partial y^2} w' + \frac{1}{2}\left(\frac{\partial w'}{\partial z}\right)^2 \Bigg]$$

(11.52)

The variational equation associated with (11.52) is simply

$$\delta \Phi = 0 \qquad (11.53)$$

As usual in the local potential technique, we have two kinds of variables, fluctuating and non fluctuating ones, these last being indicated by a superscript (o) .

The Euler Lagrange equation applied to (11.52) restores the excess balance equations (11.28 - 11.32) for mass, momentum and energy, using the a posteriori subsidiary conditions

$$\begin{aligned} n_1^{(o)} &= n_1 \\ \vartheta^{(o)} &= \vartheta \\ u'^{(o)} &= u' \\ v'^{(o)} &= v' \\ w'^{(o)} &= w' \end{aligned} \tag{11.54}$$

Again we made the same choice (11.38 - 11.44) for the disturbances and we substitute into (11.52). In the next step, we eliminate the pressure amplitude disturbance Δ in the local potential thanks to Eq. (11.20) or (11.21) and (11.38 - 11.44). The local potential becomes

$$\Phi = \int dz \Big[- \sigma \, Sc \, X^{(o)} X + s \frac{Sc}{Pr} W^{(o)} X + k^2 X^{(o)} X +$$

$$+\frac{1}{2}\left(\frac{dX}{dz}\right)^2 - s X^{(o)}\frac{dX}{dz} + s^2\Theta^{(o)}\frac{dX}{dz}$$

$$+ s N_1^S k^2 \Theta^{(o)} X + s^2\frac{d\Theta^{(o)}}{dz} X + s N_1^S \frac{d\Theta^{(o)}}{dz}\frac{dX}{dz} - \sigma P_r \Theta^{(o)}\Theta$$

$$-W^{(o)}\Theta + k^2\Theta^{(o)}\Theta + \frac{1}{2}\left(\frac{d\Theta}{dz}\right)^2 - \frac{\sigma}{k^2}\frac{dW^{(o)}}{dz}\cdot\frac{dW}{dz}$$

$$+\frac{2}{k^2}\frac{d^2W^{(o)}}{dz^2}\frac{d^2W}{dz^2} + \frac{dW^{(o)}}{dz}\cdot\frac{dW}{dz} - \frac{1}{2k^2}\left(\frac{d^2W}{dz^2}\right)^2$$

$$\left. - \sigma W^{(o)} W - \mathcal{R}_a \Theta^{(o)} W + R_{TH} X^{(o)} W + k^2 W^{(o)} W + \frac{1}{2}\left(\frac{dW}{dz}\right)^2\right] \tag{11.55}$$

The Euler Lagrange equations:

$$\frac{\delta\Phi}{\delta W} = \frac{\partial\Phi}{\partial W} - \frac{d}{dz}\frac{\partial\Phi}{\partial\left(\frac{dW}{dz}\right)} + \frac{d^2}{dz^2}\frac{\partial\Phi}{\partial\left(\frac{d^2W}{\partial z^2}\right)} \tag{11.56}$$

$$\frac{\delta\Phi}{\delta\Theta} = \frac{\partial\Phi}{\partial\Theta} - \frac{d}{dz}\frac{\partial\Phi}{\partial\left(\frac{d\Theta}{dz}\right)} + \frac{d^2}{dz^2}\frac{\partial\Phi}{\partial\left(\frac{d^2\Theta}{\partial z^2}\right)} \tag{11.57}$$

(11.58) $$\frac{\delta\Phi}{\delta X} = \frac{\partial\Phi}{\partial X} - \frac{d}{dz}\frac{\partial\Phi}{\partial\left(\frac{dX}{dz}\right)} + \frac{d^2}{dz^2}\frac{\partial\Phi}{\partial\left(\frac{d^2X}{\partial z^2}\right)}$$

applied to (11.55) together with the subsidiary conditions (11.54) restore the dispersion equations (11.45) - (11.47).

11.3.5 Eigenvalue Problem

We now have to substitute in the local potential (11.55) suitable trial functions for X, W, Θ, satisfying the boundary conditions:

(11.59) $$\Theta = 0$$

(11.60) $$W = 0$$

(11.61) $$DW = 0$$

(11.62) $$DX - sX + sN_1^s D\Theta = 0$$

for $z = 0$ and $z = 1$

Conditions (11.61) arises from the continuity equation. For W and Θ we take the already used trial functions:

(11.63) $$W = A(1 - z)^2 z^2$$

and

$$\Theta = B(1 - z)z \tag{11.64}$$

where A and B are the variational parameters. We develop X in powers of z ; since it has to satisfy the two boundary conditions (11.62), we limit ourselves to the first three terms in order to have only one independent variational parameter. Of course, as (11.62) includes the temperature amplitude disturbance, the variational parameter B will appear in X . So:

$$X = -B\left[s\left(1-\frac{s}{2}\right)z - \left(\frac{2s - s^2(1-s/2)}{2-s}\right)z^2\right] + C\left[1 + sz + \frac{s^2}{2-s}z^2\right] \tag{11.65}$$

In order to obtain (11.65), we have also substituted in (11.62) N_1^s using the linear form, $N_1^s = 1 + s\left(z - \frac{1}{2}\right)$. Substitution of the trial functions in the local potential (11.55) integration over z and finally minimization with respect to A,B and C yield the following homogeneous linear algebraic equations:

$$A\left[-\sigma\left(\frac{2}{105k^2} + \frac{1}{630}\right) + \frac{k^2}{630} + \frac{4}{105} + \frac{4}{5k^2}\right] + B\left[-\frac{\mathcal{R}a}{140} + R_{TH}f_1\right] + C\left[R_{TH}\left(\frac{1}{30} + f_2\right)\right] = 0 \tag{11.66}$$

$$A\left[\frac{S_c}{P_r} f_3 - \frac{1}{140}\right] + B\left[-\sigma\left(\frac{P_r}{30} + S_c\, f_4\right) + \frac{k^2}{30} + \frac{1}{3} + k^2 f_5 + f_6\right] \tag{11.67}$$
$$+ C\left[-\sigma S_c\, f_7 + k^2 f_7 + f_9\right] = 0$$

$$A\left[\frac{S_c}{P_r} f_{10}\right] + B\left[-\sigma S_c\, f_7 + k^2 f_{12} + f_{13}\right] \tag{11.68}$$
$$+ C\left[-\sigma\left(S_c\, f_{14} + S_c\right) + k^2 f_{14} + k^2 + f_{16}\right] = 0$$

in which f_1, f_2, f_3, ..., f_{16} are only functions of the Soret number s. The detailed expressions of these coefficients are given in table 7.

System (11.66) - (11.68) has non-trivial solutions only when the determinate of the coefficients is equal to zero. The subsequent equation of the third degree in σ has to be solved for various values of the parameters P_r, S_c, R_{TH}, $\mathcal{R}_a$, s, and k. The solutions are real or complex and the sign of the real part (σ_R) defines the stable and the unstable domain in the k-$\mathcal{R}_a$ plane (for given P_r, S_c, R_{TH}, s) limited by the neutral stability curve $\sigma_R = 0$. If the complex part (σ_I) is equal to zero, transition occurs through a stationary state.

Of course, the trial functions (11.63) (11.64)

(11.65) are as simple as possible and are only a first approximation of the actual solution. They are chosen in order to have only one independent variational parameter in each function and to carry out the calculations without excessive algebra. For this reason the numerical results given in the next paragraph are not expected to be in excellent agreement with experiments, but the probable discrepancy could be reduced by more suitable trial functions containing more variational parameters.

11.3.6 Numerical Results For $s > 0$

In this case, the more dense component migrates toward the cold plate, i.e. the upper boundary, and we expect a destabilizing effect. One of the most important features is that for $s > 0$ and for all values of the other parameters, the solutions for σ are always real. Thus, the principle of exchange of stability is valid and transition to a new steady state occurs beyond the instability point.

We want to point out that the principle of exchange of stability is found by extrapolation of the numerical results.

Some neutral stability curves ($\sigma_R = 0$) are plotted on Fig. 7.

Tables 8 to 11, give the critical Rayleigh and wave numbers for different values of R_{TH} and s, for a Prandtl number equal to 10 and a Schmidt number equal to 1000. As can be seen from these tables the critical Rayleigh number decreases when the thermal diffusion contribution increases. This is in agreement with the intuitive prediction of a destabilizing effect, which has never been observed in the experiments of Legros [1, 2, 3].

For experiments in liquid phase, one usually has $P_r \simeq 10$, and $Sc \simeq 1000$, i.e. the values used in our numerical computations. For the water-methanol mixture used by Legros et al. 3) in his experiments, $R_{TH} \simeq 40.000$ and $D'/D \simeq 2.10^{-3}$ (for a 50% in weight mixture).

From table 10, we see that the system becomes unstable for very small temperature gradient, always exceeded in the experiments reported in Ref. 3. Moreover the critical wavenumber is also decreasing when R_{TH} and s are increasing.

For realistic values of the thermal diffusion contribution, the wavelength linked to the size of the convective cells is thus very large and we expect to have only one cell of convection.

This cell could destroy the thermal diffusion gradient of concentration if the relaxation time of thermal diffusion is large with respect to the convection velocity.

The contribution of this convective movement to the total heat flux is obviously too small to be detected experimentally by measuring the temperature difference between the hot and the cold plate. So this primary instability was not seen in the experiments by Legros et al. [1, 2, 3]. We believe that this unique convection cell does, in fact, destroy the concentration gradient due to thermal diffusion and what Legros observed is a second instability point, which is the usual breakdown into the multicellular flow pattern. This could explain the fact that they never observe any destabilizing effect in systems for which $\mathfrak{s} > 0$. This unique cell is supposed to exist from the linear stability analysis. To have a complete proof we have to calculate the streamlines and the velocity profile in this cell (but clearly, this is beyond the linear stability analysis presented here) and to "record" its existence by some appropriate photographic technique. These points are now under investigation.

11.3.7 Numerical Results for $\mathfrak{s} < 0$

The more dense component migrates toward the bottom of the cell, and as it was expected, Legros observed [3] a stabilizing effect in some experiments. In this case, the calculated amplification factor σ as a function of k and $\mathcal{R}_a$ is usually complex. Then the transition beyond the critical point occurs through an oscill tory state, except for very small values of R_{TH} or $\mathfrak{s}$.

In Tables 12, 13, 14, we give the critical points in the case of overstability $(\mathcal{R}_{a\ I}^{\ cr})$ and in the case of exchange of stability $(\mathcal{R}_{a\ II}^{\ cr})$.

From Tables [12, 13, 14], we can see that the critical Rayleigh number is smaller than the one predicted wrong ly assuming the validity of the principle of exchange of stability.

A direct comparison between the numerical results presented here and experiments of Ref. 1, 2, 3 are not possible. The calculations are performed for dilute solutions and the experiments do not obey this assumption. But in concentrated solutions Legros et al. [1, 2, 3] observed an encrease in the critical temperature difference (thus in the critical Rayleigh number) in systems with a $D'/D < 0$. This increase was of the order of 25% in a mixture 90% water - 10% methanol, and from Table 13, this increase in dilute solutions is approximately 15%, thus of the good order of magnitude.

References

(Section 11)

[1] J.C. Legros, W.A. Van Hook, G. Thomaes, Chem. Phys. Letters 1, 696 (1968).

[2] J.C. Legros, W.A. Van Hook, G. Thomaes, Chem. Phys. Letters 2, 249 (1968).

[3] J.C. Legros, D. Rasse, G. Thomaes, Chem. Phys. Letters 4, 632 (1970).

[4] J.C. Legros, J.K. Platten, P. Poty, submitted to Phys. of Fluids.

[5] S. Chandrashekhar, Hydrodynamic and Hydromagnetic Stability, Oxford, Clarendon Press (1961).

[6] R.S. Schechter, D.M. Himmelblau, Phys. of Fluids, 8, 1431 (1965).

[7] J.C. Legros, W.A. Van Hook, G. Thomaes, Chem. Phys. Letters 2, 251 (1968).

[8] J.C. Legros, D. Rasse, G. Thomaes, Phisica, 57, 585 (1972).

[9] R.S. Schechter, I. Prigogine, J.R. Hamm, Phys. of Fluids March, 3 (1972).

[10] P. Glansdorff, I. Prigogine, Physica 30, 351 (1964).

[11] I. Prigogine, P. Glansdorff, Physica 31, 1242 (1965).

[12] For a general review, see for example:

a. P. Glansdorff, I.Prigogine, Thermodynamic Theory of Structure and Fluctuations, Wiley, Interscience, London, 1971.

b. R.J. Donelly, R. Herman, I. Prigogine, Non Equilibrium Thermodynamics, Variational Techniques and Stability, The University of Chicago Press (1968).

c. R.S. Schechter, The Variational Method in Engineering, Mc Graw Hill (1967).

[13] S.R. De Groot, L'Effet Soret, Thesis, North Holland Publ. Co. Amsterdam (1945).

[14] J.K. Platten, Int. J. Eng. Sci., 9 37 (1971).

[15] J.K. Platten, Int. J. Eng. Sci., 9, 855 (1971).

[16] J.K. Platten, private communication.

[17] D.R. Caldwell, J. Fluid. Mech. 161 (1970).

12. Stability of Simple Waves

12.1 Introduction

In this section we will investigate the linear stability of an isentropic flow of ideal fluids with respect to isentropic perturbations. We then have only travelling perturbations such as e.g. sound waves.

More precisely we investigate the stability of a one dimensional isentropic simple wave both for compression and rarefaction. The main feature is that we are here concerned with the stability problem for a time dependent state of reference where the kinetic method based on the normal mode analysis is pratically ruled out.

Let us recall briefly some basic properties of travelling waves. (For more information see [3] and [2]).

12.2 Sound Waves

As it is well known, for waves of infinitesimal amplitude the excess balance equations for isentropic perturbations lead to the following linear equation for the velocity potential:

(12.1) $$\partial_t^2 \Phi - c^2 (\Phi_{,i})_{,i} = 0 \qquad \begin{cases} c^2 = \left(\dfrac{\partial p}{\partial \rho}\right)_s \\ u_i = \delta v_i = \dfrac{\partial \Phi}{\partial x_i} = \Phi_{,i} \end{cases}$$

The unperturbed reference state is then stable with respect to sound waves. However the situation changes completely when the basic reference state is no longer homogeneous. The wave equation becomes then non-linear.

12.3 Compression and Rarefaction Waves–Riemann Invariants

Special cases of one dimensional gas flow are the compression and the rarefaction waves. To produce them one has to move a piston in or out of a pipe. In both cases we suppose that the disturbance produced by the piston's motion propagates to the right. (For more details see [3]).

Assuming the flow to be isentropic, a combination of the continuity equation and that of Euler's gives us:

(12.2) $$\left[\frac{\partial v}{\partial t} \pm \frac{1}{\rho c}\frac{\partial p}{\partial t}\right] + (v \pm c)\left[\frac{\partial v}{\partial x} \pm \frac{1}{\rho c}\frac{\partial p}{\partial x}\right] = 0$$

Introducing the notion of <u>characteristics</u> we may rewrite (12.2) as:

(12.3) $$dv + \frac{1}{\rho c}\, dp = dJ_+ = 0 \quad \text{on } C_+\left(\frac{dx}{dt} = v + c\right)$$

$$dv - \frac{1}{\rho c}\, dp = dJ_- = 0 \quad \text{on } C_-\left(\frac{dx}{dt} = v - c\right) \tag{12.4}$$

where the operator d denotes derivation along the corresponding characteristic C_-. The functions J_+ and J_- introduced here are called the Riemann invariants. In the case of a perfect gas, using the adiabatic equation we get for the latter:

$$J_\pm = v \pm \frac{2c}{\gamma - 1} \tag{12.5}$$

Let us recall that:

1°. J_- reduces to a simple constant in all the x-t diagram

2°. the C_+ characteristic are straight lines [3] .

Along these we get successively:

$$x = \left[v + c\left(v\right)\right]t + \varphi(v) \quad \text{on } C_+ \tag{12.6}$$

$$v = f\left\{x - \left[v + c\left(v\right)\right]t\right\} \tag{12.7}$$

$$c = g\left\{x - \left[v + c\left(v\right)\right]t\right\} \tag{12.8}$$

Waves of finite amplitude defined by equations (12.6), (12.7) and (12.8) are called simple waves.

3°. For a rarefaction wave one has:

$$(12.9)\qquad \frac{\partial v}{\partial x} > 0\,,\quad \frac{\partial p}{\partial x} > 0\,,\quad \frac{\partial \rho}{\partial x} > 0\,,\quad \frac{\partial c}{\partial x} > 0$$

The C_+ characteristics originating from the piston form a family of divergent straight lines.

4°. For a compression wave one has:

$$(12.10)\qquad \frac{\partial v}{\partial x} < 0\,,\quad \frac{\partial p}{\partial x} < 0\,,\quad \frac{\partial \rho}{\partial x} < 0\,,\quad \frac{\partial c}{\partial x} < 0$$

The C_+ characteristics form a family of convergent straight lines which finally intersect. As each characteristic corresponds to a well defined velocity v , this means that the hydrodynamic velocity is no longer a uniform function and we have then formation of a shock wave.

12.4. Small Disturbances of Travelling Waves

Using equations (12.3) and (12.4) and neglecting higher order terms in the fluctuations we may write:

$$(12.11)\qquad \delta J_{\pm} = u \pm \frac{1}{\rho c}\,\varpi$$

The two perturbations δJ_+ and δJ_- are then propagated respectively and independently along the C_+ and C_- characteristics. Let us first suppose:

$$(12.12)\qquad \delta J_- = 0$$

Equation (12.12) leads then to the following equalities for the δJ_+ perturbations

$$\varpi = \rho c u \tag{12.13}$$

$$\delta\rho = \frac{\rho}{c} u \tag{12.14}$$

$$\delta T = \vartheta = \frac{\alpha T c}{c_p} u \tag{12.15}$$

Taking now the opposite case:

$$\delta J_+ = 0$$

we are led to:

$$\varpi = -\rho c u \ ; \quad \delta\rho = -\frac{\rho}{c} u \ ; \quad \vartheta = -\frac{\alpha T c}{c_p} u \tag{12.16}$$

12.5 Instability of the Simple Compression Wave

The failure of the hydrodynamic stability criterion $P\left[\delta E_{kin}\right] \leqslant 0$ is used below as a sufficient condition for instability, and will enable us e.g. to prove the instability of a compression wave.

Taking (12.13) and (12.14) into account and putting $\tau^2 = 1, p_{ij} = 0$, we get along the C_+ characteristics the following stability condition (for more details see [1]):

$$P\left[\delta E_{kin}\right] = -\int_{x_p}^{\infty} u^2 \left[\frac{1}{2} \frac{\partial(\rho v + \rho c)}{\partial x} + \frac{\rho}{c}(v+c)\frac{\partial v}{\partial x}\right] dx < 0$$

(12.17)

One sees immediately using the inequalities (12.10) that the above condition is violated. The simple compression wave is therefore unstable.

Let us stress on the fact that even in this simplified model the above instability is still situated at a finite distance of equilibrium, as waves of finite amplitude have to build up first, for the corresponding gradients to have non negligible values. However a complete treatement implies to investigate the role of the dissipative effects on the perturbations even when the state of reference is a simple wave. The above question is actually under study.

12.6 Stability of the Simple Rarefaction Wave

Let us stress that the sign reversal of the gradients (compare (12.9) to (12.10)), in this case does not permit to conclude the stability, the hydrodynamic stability criterion being incomplete as δE_{kin} is only a semi-definite form.

We now have to use the general stability criterion based on the sign of the excess entropy production (see section 6).

The appropriate form of $P\left[\delta Z\right]$ for this particular problem is obtained using the isentropic one dimensional assumptions along with the equations (12.13), (12.14), (12.15) and (12.16).

After elementary manipulations (cf. [1]) we get for the generalized entropy source:

$$\sigma_{\pm} = \frac{u^2}{T}\left\{\left[\frac{\alpha}{c_p}(2v \pm c) + \frac{1}{c^2}(v \pm c)\right]\frac{\partial p}{\partial x} + \left(2 + \frac{\alpha c^2}{c_p} \pm \frac{v}{c}\right)\rho\frac{\partial v}{\partial x} - \frac{\rho v}{c}\frac{\partial c}{\partial x}\right\} \tag{12.18}$$

Using the adiabatic equation for a perfect gas to establish links among the three gradients we finally derive from (12.5):

$$P_{+}\left[\delta Z\right] = \int_{x_p}^{\infty}\frac{\rho u^2}{(\gamma-1)T}\left[(3\gamma+1)\frac{v}{c} + u\gamma + 2\right]\frac{\partial c}{\partial x}\,dx \geqslant 0 \quad (\text{on } C_{+}) \tag{12.19}$$

$$P\left[\delta Z\right] \quad \int_{x_p}^{\infty}\frac{\rho u^2}{(\gamma-1)T}\left[3(\gamma-1)\frac{v}{c} + 2\right]\frac{\partial c}{\partial x}\,dx \geqslant 0 \quad (\text{on } C_{-}) \tag{12.20}$$

The inequalities (12.19) and (12.20) are certainly satisfied for the subsonic region ($|V| < c$) and for usual values of γ as it can easily be verified (For exceptional cases see [4] and [5]).

<u>The rarefaction simple wave is then a stable time-dependent process in the subsonic region.</u>

References

(Section 12)

[1] P. Glansdorff and I. Prigogine, Thermodynamic Theory of Structure, Stability and Fluctuations, Wiley, Interscience, London, 1971.

[2] Ya. B. Zeldovich and Yu. P. Raizer, Physics of Shock Waves and High temperature - Hydrodynamic Phenomena, Vol. 1, Ed. W. D. Hayes and R.F. Probstein, Academic Press, New York and London, 1966.

[3] L. Landau and E. M. Lifshitz, Fluid Mechanics, Pergamon Press, London, 1959.

[4] G.D. Kahl and D.C. Mylin, Phys. Fluids, 12, 11 (1969).

[5] L. Landau, Collected Papers, "On a Study of the Detonation of Condensed Explosives", Pergamon Press, New York, London, p. 425, 1965.

13. Stability of Plane Poiseuille Flow

13.1 Introduction

In this section we use the local potential technique to investigate the linear stability of a steady plane Poiseuille flow of an incompressible fluid.

The dimensionless parameter characterizing the problem is presently the Reynolds number:

$$\mathcal{R}_e = \frac{U^+ h}{\nu}$$

where U^+ is some reference velocity, usually chosen as half the maximum velocity for the above flow.

Before proceeding any further let us recall briefly the eigenvalue approach to this problem. For more details see [2].

13.2 The Eigenvalue Problem for Hydrodynamic Stability

Starting from the excess balance equations for mass and momentum in the absence of external forces we obtain for an incompressible fluid, the set of linearized perturbation equations:

$$u_{i,i} = 0 \tag{13.1}$$

(13.2) $$\partial_t u_i = -\bar{U}_j u_{i,j} - u_j \bar{U}_{i,j} - \varpi_{,i} + (\mathcal{R}e)^{-1} (u_{i,j})_{,j}$$

($u_i = \delta v_i$; $\varpi_i = \delta p$; $\bar{U}_i$: velocity of the state of reference;

$$u_{i,j} = \frac{\partial u_i}{\partial x_j})$$

The different variables have been reduced, using one or another appropriate set of reducing factors.

Now, writing the perturbations as

(13.3)
$$\begin{aligned} u_1 = u &= U(z) \quad e^{i\alpha(x-ct)} \\ u_2 = w &= W(z) \quad e^{i\alpha(x-ct)} \\ \varpi &= \Pi(z) \quad e^{i\alpha(x-ct)} \end{aligned}$$

replacing in (13.1) and (13.2) and eliminating U and Π, we get the well known <u>Orr-Sommerfeld equation</u> [2] :

(13.4)
$$\left(D^2 - \alpha^2\right)^2 W = i\alpha \mathcal{R}_e \left[\left(\bar{U} - c\right)\left(D^2 - \alpha^2\right) W - \left(D^2 \bar{U}\right) W\right]$$
$$\left(D = \frac{d}{dz} ; \qquad c = c_r + i\, c_i\right)$$

The boundary conditions here adopted are:

(13.4') $$W = DW = 0 \qquad \left(\text{for } z = z_1 ; z = z_2\right)$$

As usual, the relation between α and $\mathcal{R}e$ for $c_i = 0$

(marginal state) should provide us with the curve of neutral stability, from which the critical Reynolds number can be deduced. However due to the non-self adjoint character of the Orr-Sommerfeld equation, the solution to this problem is rather complicated and the use of the local potential technique is again justified (see [1]).

13.3 The Excess Local Potential for Hydrodynamic Stability

Considering the perturbations' evolution as a special case of macroscopic motion, we can construct thanks to the excess balance equations an excess local potential.

In order to illustrate the generality of the local potential technique, we will first derive the suitable expression of the excess local potential for a laminar flow of a non isothermal fluid (transverse temperature gradient). Once the latter is established, we will then confine ourselves for numerical calculations to the more elementary Poiseuille (isothermal) plane flow. (see [1]).

Using again appropriate reducing, factors the excess balance equations for momentum and energy give for the plane Poiseuille flow:

$$\partial_t w = - \mathcal{R}_e \overline{U} w_{,x} - \varpi_{,z} + \mathcal{R}_a \vartheta + \nabla^2 w \qquad (13.5)$$

$$\partial_t u = - \mathcal{R}_e \overline{U} u_{,x} - \varpi_{,x} - \mathcal{R}_a w \overline{U}_{,z} + \nabla^2 u \qquad (13.6)$$

$$\mathcal{P}_r \ \partial_t \vartheta = - \ \mathcal{P}_e \ \bar{U} \ \vartheta_{,x} + w + \nabla^2 \vartheta$$

(13.7) $$\left(u = \delta v_x \ ; \quad w = \delta v_z \ ; \quad \vartheta = \delta T ; \mathcal{P}_r = \frac{\nu}{\chi} ; \ \mathcal{P}_e = \frac{U^+ h}{\chi} \right)$$

To construct the corresponding local potential we multiply the two sides of (13.5), (13.6) and (13.7) by the small increments $-\Delta w - \Delta u$ and $-\Delta \vartheta$ respectively and add. Using then the usual method, we obtain a time dependent excess local potential in the form (see [1]):

$$\Psi = \Psi \left(u, \overset{\circ}{u}, w, \overset{\circ}{w}, \vartheta, \overset{\circ}{\vartheta}, \dot{\varpi}, \mathcal{R}_e, \mathcal{R}_a, \mathcal{P}_r \right)$$

Let us assume the following expressions for the disturbance (*)

(13.8) $$\begin{aligned} w &= W(z) \, e^{i\alpha x} \, e^{\sigma t} \\ u &= \frac{i}{\alpha} DW(z) \, e^{i\alpha x} \, e^{\sigma t} \\ \vartheta &= \Theta(z) \, e^{i\alpha x} \, e^{\sigma t} \\ \varpi &= \Pi(z) \, e^{i\alpha x} \, e^{\sigma t} \end{aligned}$$

(*) The functions adopted having a period T along the x -axis, the integrations over this axis are to be carried out over nT where n is an integer.

where σ denotes a reduced eigenvalue. Using these and eliminating Π_0 , we get a time independent excess local potential depending only on two unknown functions, $W(z)$ and $\Theta(z)$ under the form:

$$\Psi = \int_{z_1}^{z_2} \Big\{ \left[\alpha^2 + \sigma + i\alpha \mathcal{R}_e \bar{U}\right]\left[W^\circ W + \frac{1}{\alpha^2}(DW^\circ)(DW)\right]$$

$$+ \left[\alpha^2 + \mathcal{P}_r \left(\sigma + i\alpha \mathcal{R}_e \bar{U}\right)\right]\Theta^\circ \Theta + \frac{1}{2}\left[(DW)^2 + (D\Theta)^2\right]$$

$$- \mathcal{R}_a \Theta^\circ W - W^\circ \Theta - \frac{1}{2\alpha^2}(D^2 W)^2 + \frac{2}{\alpha^2}(D^2 W^\circ)(D^2 W)$$

$$- \frac{i\mathcal{R}_e}{\alpha}(D\bar{U})(DW)W^\circ \Big\} dz \qquad (13.9)$$

Let us observe that this local potential involves separate particular cases. The first corresponds to $\mathcal{R}_a = 0$ (isothermal Poiseuille flow), the second to $\mathcal{R}_e = 0$ (Bénard problem). We shall only consider the first case.

13.4 The Approximative Critical Reynolds Number for the Plane Poiseuille Flow

We obtain the suitable local potential for this problem by putting in (13.9) $\sigma = -i\alpha\mathcal{R}_e c$ (as the extremal has to yield the Orr–Sommerfeld equation), $\Theta = 0$ and multiplying it by α^2. This yields:

$$\Phi = \int_{-1}^{+1}\Big\{\Big[\alpha^4 + i\alpha^3 \mathcal{R}_e\ (\bar{U} - c)\Big]\Big[\overset{\circ}{W} W + \frac{1}{\alpha^2}(D\overset{\circ}{W})(DW)\Big]$$

(13.10) $$+ \frac{\alpha^2}{2}(DW)^2 - \frac{1}{2}(D^2W)^2 + 2(D^2\overset{\circ}{W})(D^2W)$$

$$- i\alpha \mathcal{R}_e\ (D\bar{U})(DW)\overset{\circ}{W}\Big\}dz$$

Let us introduce in (13.10) the sequence of order n of the trial functions:

(13.11) $$W_n = a_1\varphi_1 + \ldots + a_n\varphi_n;\quad \overset{\circ}{W}_n = \overset{\circ}{a}_1\varphi_1 + \ldots + \overset{\circ}{a}_n\varphi_n$$

the φ_i satisfying the boundary conditions (13.4').
Using now, the classical minimization procedure we are led to:

(13.12) $$\mathrm{Det}\left|A_{k\ell} - c\,B_{k\ell}\right| = 0 \qquad (k, \ell = 0, \ldots n)$$

The eigenvalue problem for c corresponding to (13.12), is then solved on a computer. An approximate curve of neutral stability is derived (see fig. 1). Platten has obtained in this way the good approximation (see [3]):

$$5600 < (\mathcal{R}_e)_c < 5900$$

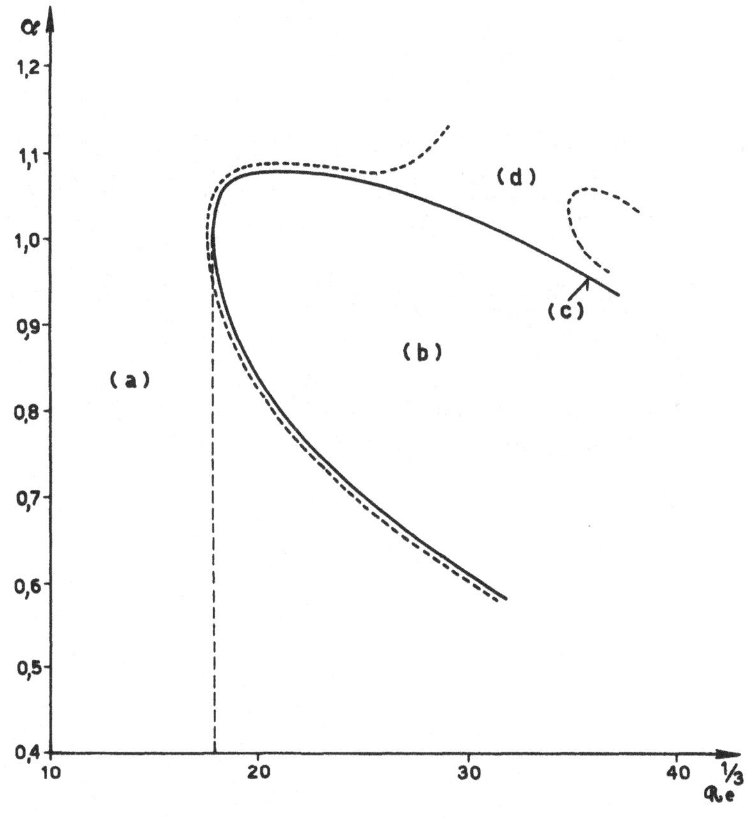

Fig. 1.

(a) – Stable region

(b) – Unstable region

(c) – Neutral stability

(d) – Indetermination due to the numerical approximation.

References

(Section 13)

[1] P. Glansdorff and I. Prigogine, Thermodynamic Theory of Structure, Stability and Fluctuations, Wiley, Interscience, London, 1971.

[2] C.C. Lin, The Theory of Hydrodynamic Stability, Cambridge University Press, London, 1955.

[3] J. Platten, Thèse de doctorat, Chimie-Physique, Université de Bruxelles, 1970.

Glossary of Principal Symbols

Greek symbols

α	expansion coefficient
α_n	parameters in test functions
γ	expansion coefficient due to concentration species symbol
δ	variational symbol
Δ	finite increment
Δ_γ	diffusion velocity of γ
∇	gradient operator
∇^2	Laplacian operator
$\varepsilon; \tau$:	weighting function
$\eta=\nu\rho$	dynamic viscosity
ϑ	temperature perturbation
χ	thermal diffusivity
λ	thermal conductivity;wave length
μ_γ	chemical potential per unit mass of γ
$\nu_{\gamma\rho}$	stoechiometric coefficient of a chemical reaction ρ
ν	kinematic viscosity frequency
ξ_ρ	degree of advancement or extend of reaction ρ
$\varpi=\delta p$	hydrostatic pressure perturbation
ρ	density

σ	non-dimensional eigenvalue
χ	isothermal compressibility
Ω	surface
$\omega = \omega_r + i\,\omega_i$	eigenvalue of a complex normal mode

Latin symbols

A_ρ	Affinity of the chemical reaction
$c_v(c_p)$	specific heat at constant volume (pressure)
c_γ	molar concentration of γ
c	sound velocity
D_γ	diffusion coefficient of γ
d_t	time derivative
$D = \frac{d}{dz}$	(z = reduced height)
E	internal energy
e	internal energy per unit mass
$,i = \partial/\partial x_i$	
j	density of flow
k	Boltzmann's constant
$L_{\alpha\beta}$	phenomenological coefficient
M_γ	molar mass of γ
N_γ	mass fraction of γ
$P[I]$	global production of I per unit time
p_{ij}	non-equilibrium part of the pressure tensor
$\mathcal{S}$	density of the entropy flow

U	total energy
$u = \delta v_i \, (\delta v_z = w)$	velocity perturbation
V	total volume
$\mathbf{v}$	barycentric velocity
$v \, (\rho^{-1})$	specific volume
W	amplitude factor of the perturbation w
$\mathbf{W}$	heat flow
w_ρ	rate of the $\rho^{\underline{th}}$ chemical reaction
$+$	reference state symbol

TABLE 1

System $C_6H_6 - CCl_4$
Thickness of the fluid layer : 0.9 mm ± 0.03 mm

X(mass fraction)$_{CCl_4}$	ΔT^{cr} (°C.)
1.00	1.63
0.83	1.74
0.80	1.68
0.70	1.65
0.67	1.87
0.63	1.85
0.45	2.08
0.28	2.15
0.00	2.71
0.00	2.53

TABLE 2
System $C_6H_5Cl - CCl_4$
Thickness of the fluid layer : 1.19 mm ± 0.03 mm

X_{CCl_4}	ΔT^{cr}
0.00	5.20
0.25	4.50
0.52	3.70
0.74	3.50
1.00	3.17

TABLE 3
System $C_2H_2Cl_4 - C_2H_2Br_4$
Thickness of the fluid layer : 1.19 mm ± 0.03 mm

$X_{C_2H_2Cl_4}$	ΔT^{cr}
1.00	6.00
0.50	7.80
0.00	16.75

TABLE 4
System $C_6H_{12} - CCl_4$
Thickness of the fluid layer : 1.29 mm ±0.03 mm

$X_{C_6H_{12}}$	ΔT^{cr}
1.00	5.65
0.50	3.62
0.00	2.80

TABLE 5
System $H_2O - CH_3OH$
Thickness of the fluid layer : 3.13 mm ±0.03 mm

	A^+	B^+
1.00		2.74
0.90	3.20	2.81
0.80	2.56	2.32
0.60		1.71
0.30		1.02
0.00		0.65

TABLE 6

System $H_2O - C_2H_5OH$

Thickness of the fluid layer : 3.13 mm ±0.03 mm

X_{H_2O}	A	B
1.00		2.74
0.95	3.58	3.03
0.90	3.65	3.00
0.85	3.30	2.83
0.80		2.66
0.75		2.41
0.40		1.43
0.00		0.85

TABLE 7

The f polynoms

$$f_1 = -\frac{1}{60} E + \frac{1}{105} D$$

$$f_2 = \frac{1}{60} s + \frac{1}{105} F$$

$$f_3 = -E s \frac{1}{60} + D s \frac{1}{105}$$

$$f_4 = \frac{1}{3} E^2 - \frac{1}{2} DE + \frac{1}{5} D^2$$

$$F_5 = \frac{1}{3}E^2 - \frac{1}{2}DE + \frac{1}{5}D^2 - \frac{1}{12}Es + \frac{1}{20}Ds - \frac{1}{120}Es^2 + \frac{1}{120}Ds^2$$

$$F_6 = E^2 - 2DE + \frac{4}{3}D^2 - \frac{1}{2}E^2 s + DEs - \frac{1}{2}D^2 s \; \frac{1}{3}Ds + \frac{1}{6}Es^2 - \frac{1}{6}Ds^2$$

$$F_7 = -\frac{1}{2}E + \frac{1}{3}D - \frac{1}{3}Es + \frac{1}{4}Ds - \frac{1}{4}EF + \frac{1}{5}FD$$

$$F_8 = F_7$$

$$F_9 = -EF + \frac{4}{3}DF + \frac{1}{2}Es^2 - \frac{2}{3}Ds^2 + \frac{1}{3}EFs - \frac{1}{2}DFs$$

$$F_{10} = \frac{1}{30}s + \frac{1}{60}s^2 + \frac{1}{105}Fs$$

$$F_{11} = -\frac{1}{2}E + \frac{1}{3}D - \frac{1}{3}Es + \frac{1}{4}Ds - \frac{1}{4}EF + \frac{1}{5}FD$$

$$F_{12} = \frac{1}{2}E + \frac{1}{3}D - \frac{1}{3}Es + \frac{1}{4}Ds - \frac{1}{4}EF + \frac{1}{5}FD + \frac{1}{6}s$$

$$+ \frac{1}{12}s^2 + \frac{1}{120}s^3 + \frac{1}{20}Fs + \frac{1}{120}Fs^2$$

$$F_{13} = -Es - EF + \frac{4}{3}DF + Ds + \frac{1}{2}Es^2 + \frac{2}{3}FEs - \frac{1}{2}FDs$$

$$-\frac{1}{3}Ds^2 - \frac{1}{3}Fs - \frac{1}{6}s^3 - \frac{1}{6}Fs^2$$

$$F_{14} = s + \frac{2}{3}F + \frac{1}{3}s^2 + \frac{1}{2}sF + \frac{1}{5}F^2$$

$$F_{15} = s + \frac{2}{3} F + \frac{1}{3} s^2 + \frac{1}{2} sF + \frac{1}{5} F^2 = F_{14}$$

$$F_{16} = \frac{4}{3} F^2 + Fs - \frac{1}{2} s^3 - F s^2 - \frac{1}{2} F^2 s$$

with : $$D = \frac{2s - s^2\left(1 - \frac{s}{2}\right)}{2 - s}$$

$$E = s\left(1 - \frac{s}{2}\right)$$

$$F = \frac{s^2}{2 - s}$$

TABLE 8

Critical Rayleigh number and critical wave number for $R_{TH} = 1000$ as a function of s

s	$\mathcal{R}a^{cr}$	k^{cr}
10^{-10}	1750	3.1
10^{-9}	1750	3.1
10^{-8}	1750	3.1
10^{-7}	1750	3.1
10^{-6}	1750	3.1
10^{-5}	1749	3.1
10^{-4}	1735	3.1
10^{-3}	1601	3.1
2.10^{-3}	1443	2.9
3.10^{-3}	1281	2.7
4.10^{-3}	1108	2.7
5.10^{-3}	919	2.5
6.10^{-3}	703	2.1
7.10^{-3}	404	1.3
8.10^{-3}	0	0

TABLE 9

Critical Rayleigh number and critical wave number for $R_{TH} = 10000$ as a function of δ

δ	$\mathcal{R}a^{cr}$	k^{cr}
10^{-10}	1750	3.1
10^{-9}	1750	3.1
10^{-8}	1750	3.1
10^{-7}	1750	3.1
10^{-6}	1749	3.1
10^{-5}	1735	3.1
10^{-4}	1601	3.1
2.10^{-4}	1443	2.9
3.10^{-4}	1281	2.9
4.10^{-4}	1108	2.7
5.10^{-4}	920	2.5
6.10^{-4}	703	2.1
7.10^{-4}	405	1.3
8.10^{-4}	0	0

TABLE 10

Critical Rayleigh number and critical wave number for $R_{TH} = 40\,000$ as a function of δ

δ	$\mathcal{R}a^{cr}$	k^{cr}
10^{-10}	1750	3.1
10^{-9}	1750	3.1
10^{-8}	1750	3.1
10^{-7}	1749	3.1
10^{-6}	1744	3.1
10^{-5}	1690	3.1
10^{-4}	1108	2.7
2.10^{-4}	0	0

TABLE 11

Critical Rayleigh number and critical wave number for $R_{TH} = 100\,000$ as a function of ϑ

ϑ	$\mathcal{R}a^{cr}$	k^{cr}
10^{-10}	1750	3.1
10^{-9}	1750	3.1
10^{-8}	1750	3.1
10^{-7}	1749	3.1
10^{-6}	1735	3.1
2.10^{-6}	1720	3.1
3.10^{-6}	1705	3.1
4.10^{-6}	1690	3.1
5.10^{-6}	1675	3.1
6.10^{-6}	1660	3.1
7.10^{-6}	1645	3.1
8.10^{-6}	1631	3.1
9.10^{-6}	1616	3.1
10^{-5}	1601	3.1
2.10^{-5}	1443	2.9
3.10^{-5}	1281	2.7
4.10^{-5}	1108	2.7
5.10^{-5}	920	2.5
6.10^{-5}	703	2.1
7.10^{-5}	405	1.3
8.10^{-5}	0	0

TABLE 12

Critical Rayleigh number and critical wave number for $R_{TH} = 10000$ as a function of ϑ (Oscillatory solutions = I, non oscillatory solutions = II).

ϑ	$\mathcal{R}a_{I}^{cr}$	k_{I}^{cr}	$\mathcal{R}a_{II}^{cr}$	k_{II}^{cr}
$-\ .10^{-9}$	1759.1	3.10	1750.0	3.11
$-\ .10^{-8}$	1759.1	3.10	1750.0	3.12
$-\ .10^{-7}$	1759.1	3.10	1750.0	3.10
$-\ .10^{-6}$	1759.1	3.10	1751.5	3.11
$-\ .10^{-5}$	1759.2	3.10	1764.8	3.12
$-\ .10^{-4}$	1760.0	3.10	1896.3	3.19
$-\ .10^{-3}$	1768.2	3.10	3096.2	3.73
$-\ .10^{-2}$	1846.4	3.11	12689.0	5.23
-2.10^{-2}	1925.9	3.11	22083.7	6.06
-3.10^{-2}	1997.7	3.10	30989.0	6.08
-4.10^{-2}	2062.0	3.11	39050.8	6.86
-5.10^{-2}	2119.5	3.11	46593.1	7.14
-6.10^{-2}	2170.7	3.11	53789.0	7.09
-7.10^{-2}	2216.4	3.12	60189.0	7.29
-8.10^{-2}	2257.7	3.11	65989.0	7.59
-9.10^{-2}	2295.3	3.12	70954.6	7.93
$-\ .10^{-1}$	2330.4	3.11	75398.0	8.08
-2.10^{-1}	2728.5	3.13	91003.0	9.13
-3.10^{-1}	3567.5	3.13	> 200000	
-4.10^{-1}	4992.3	3.12	> 200000	
-5.10^{-1}	6995.1	3.10	> 200000	
-6.10^{-1}	9361.4	3.09	> 200000	
-7.10^{-1}	12112.3	3.01	> 200000	
-8.10^{-1}	12120.2	2.94	> 200000	
-9.10^{-1}	18315.1	2.90	> 200000	
-1	21644.6	2.84	> 200000	
-10	> 200000		> 200000	

TABLE 13

Critical Rayleigh number and critical wave number for R_{TH} =40 000 as a function of δ (Oscillatory solutions =I, non oscillatory solutions =II).

δ	$\mathcal{R}a_{I}^{cr}$	k_{I}^{cr}	$\mathcal{R}a_{II}^{cr}$	k_{II}^{cr}
$-\ .10^{-9}$	1759.1	3.10	1750.0	3.11
$-\ .10^{-8}$	1759.1	3.10	1750.1	3.11
$-\ .10^{-7}$	1759.1	3.10	1750.6	3.11
$-\ .10^{-6}$	1759.1	3.11	1756.0	3.11
$-\ .10^{-5}$	1759.5	3.10	1809.0	3.15
$-\ .10^{-4}$	1762.7	3.11	2315.4	3.41
$-\ .10^{-3}$	1795.6	3.10	6589.0	4.25
$-\ .10^{-2}$	2120.1	3.11	40357.0	6.86
-2.10^{-2}	2472.8	3.11	74600.2	7.77
-3.10^{-2}	2816.6	3.12	107089.0	8.10
-4.10^{-2}	3151.6	3.13	137689.0	8.63
-5.10^{-2}	3478.1	3.13	166364.1	9.14
-6.10^{-2}	3796.5	3.14	193089.0	9.12
-7.10^{-2}	4107.7	3.15	>200000	
-8.10^{-2}	4412.9	3.16	>200000	
-9.10^{-2}	4713.4	3.17	>200000	
$-\ .10^{-1}$	5010.8	3.18	>200000	
-2.10^{-1}	8181.7	3.25	>200000	
-3.10^{-1}	12625.9	3.31	>200000	
-4.10^{-1}	18944.5	3.37	>200000	
-5.10^{-1}	27140.4	3.40	>200000	
-6.10^{-1}	36960.5	3.40	>200000	
-7.10^{-1}	48076.1	3.31	>200000	
-8.10^{-1}	60173.2	3.24	>200000	
-9.10^{-1}	72991.4	3.15	>200000	
-1	86331.7	3.10	>200000	

TABLE 14

Critical Rayleigh number and critical wave number for $R_{TH} = 100\,000$ as a function of $\mathfrak{s}$ (Oscillatory solutions =I, non oscillatory solutions =II).

$\mathfrak{s}$	$\mathcal{R}a_I^{cr}$	k_I^{cr}	$\mathcal{R}a_{II}^{cr}$	k_{II}^{cr}
$-.10^{-9}$	1759.1	3.10	1750.0	3.11
$-.10^{-8}$	1759.1	3.10	1750.2	3.10
$-.10^{-7}$	1759.1	3.10	1751.5	3.11
$-.10^{-6}$	1759.2	3.10	1764.8	3.12
$-.10^{-5}$	1760.0	3.10	1896.3	3.19
$-.10^{-4}$	1768.2	3.10	3095.3	3.72
$-.10^{-3}$	1850.3	3.11	12708.2	5.35
$-.10^{-2}$	2667.5	3.12	92082.0	8.09
-2.10^{-2}	3566.2	3.14	174043.0	9.14
-3.10^{-2}	4454.1	3.15	> 200000	
-4.10^{-2}	5330.1	3.18	> 200000	
-5.10^{-2}	6194.1	3.20	> 200000	
-6.10^{-2}	7046.5	3.21	> 200000	
-7.10^{-2}	7889.0	3.16	> 200000	
-8.10^{-2}	8720.7	3.25	> 200000	
-9.10^{-2}	9546.4	3.26	> 200000	
$-.10^{-1}$	10367.8	3.29	> 200000	
-2.10^{-1}	19089.0	3.20	> 200000	
-3.10^{-1}	30727.7	3.68	> 200000	
-4.10^{-1}	46833.2	3.83	> 200000	
-5.10^{-1}	67496.7	3.93	> 200000	
-6.10^{-1}	92146.7	3.96	> 200000	
-7.10^{-1}	119994.2	3.94	> 200000	
-8.10^{-1}	150272.2	3.80	> 200000	
-9.10^{-1}	182338.8	3.70	> 200000	
-1	200000			

Contents

Printed in the United States
By Bookmasters